Probabilistic Logics and Probabilistic Networks

SYNTHESE LIBRARY

STUDIES IN EPISTEMOLOGY, LOGIC, METHODOLOGY, AND PHILOSOPHY OF SCIENCE

VOLUME 350

For further volumes:
http://www.springer.com/series/6607

Probabilistic Logics and Probabilistic Networks

by

Rolf Haenni
Bern University of Applied Sciences, Switzerland

Jan-Willem Romeijn
University of Groningen, The Netherlands

Gregory Wheeler
Universidade Nova de Lisboa, Portugal

and

Jon Williamson
University of Kent, UK

 Springer

Prof. Rolf Haenni
Bern University of Applied Sciences
Department of Engineering and
Information Technology
Quellgasse 21
CH-2501 Biel
Switzerland
rolf.haenni@bfh.ch

Prof. Jan-Willem Romeijn
University of Groningen
Faculty of Philosophy
Oude Boteringestraat 52
9712 GL Groningen
Netherlands
J.W.Romeijn@rug.nl

Dr. Gregory Wheeler
Universidade Nova de Lisboa
New University of Lisbon
Quinta da Torre
2829-516 Caparica
Portugal
grw@fct.unl.pt

Prof. Jon Williamson
University of Kent
School of European Culture & Languages
Sec. Philosophy
Cornwallis North West
CT2 7NF Canterbury, Kent
United Kingdom
j.williamson@kent.ac.uk

ISBN 978-94-007-3443-2   ISBN 978-94-007-0008-6 (eBook)
DOI 10.1007/978-94-007-0008-6
Springer Dordrecht Heidelberg London New York

Springer is part of Springer Science+Business Media (www.springer.com)

Chaos *umpire sits,*
And by decision more embroils the fray
By which he reigns: next him high arbiter
Chance *governs all.*
    *(*John Milton, *Paradise Lost)*

    Chance, too, which seems to rush along
with slack reins, is bridled and governed by
law.
    *(*Boethius, *The Consolation of Philosophy)*

    Er glaubte nämlich, die Erkenntnis jeder
Kleinigkeit, also zum Beispiel auch eines sich
drehenden Kreisels, genüge zur Erkenntnis
des Allgemeinen. Darum beschäftigte er sich
nicht mit den großen Problemen, das schien
ihm unökonomisch. War die kleinste
Kleinigkeit wirklich erkannt, dann war alles
erkannt, deshalb beschäftigte er sich nur mit
dem sich drehenden Kreisel.
    *(*Kafka, *Der Kreisel)*[1]

Lest men suspect your tale untrue,
Keep probability in view.
    *(*John Gray, *The Painter who pleased*
Nobody and Everybody)

---

[1] For he believed that the understanding of any trifle, so for example of a spinning top, would suffice for the understanding of everything. This is why he did not concern himself with the big questions, which seemed uneconomical to him. If the smallest detail was truly understood, then so was everything, hence he only busied himself with the spinning top.

Preface

While in principle probabilistic logics might be applied to solve a range of problems, in practice they are rarely applied at present. This is perhaps because they seem disparate, complicated, and computationally intractable. However, we shall argue in this programmatic book that several approaches to probabilistic logic fit into a simple unifying framework: logically complex evidence can be used to associate probability intervals or probabilities with sentences.

Specifically, we show in Part I that there is a natural way to present a question posed in probabilistic logic, and that various inferential procedures provide semantics for that question: the standard probabilistic semantics (which takes probability functions as models), probabilistic argumentation (which considers the probability of a hypothesis being a logical consequence of the available evidence), evidential probability (which handles reference classes and frequency data), classical statistical inference (in particular the fiducial argument), Bayesian statistical inference (which ascribes probabilities to statistical hypotheses), and objective Bayesian epistemology (which determines appropriate degrees of belief on the basis of available evidence).

Further, we argue, there is the potential to develop computationally feasible methods to mesh with this framework. In particular, we show in Part II how credal and Bayesian networks can naturally be applied as a calculus for probabilistic logic. The probabilistic network itself depends upon the chosen semantics, but once the network is constructed, common machinery can be applied to generate answers to the fundamental question introduced in Part I.

Bern,   *Rolf Haenni*
Groningen,   *Jan-Willem Romeijn*
Lisbon,   *Gregory Wheeler*
Canterbury,   *Jon Williamson*
April 2009

Acknowledgements

This research was undertaken as a part of the *progicnet* international network on Probabilistic Logic and Probabilistic Networks. We are very grateful to The Leverhulme Trust for financial support.

Contents

Part I
Probabilistic Logics

Chapter 1
Introduction

1.1 The Fundamental Question of Probabilistic Logic

In a non-probabilistic logic, the fundamental question of interest is whether a proposition ψ is entailed by premise propositions $\varphi_1, \ldots, \varphi_n$:

$$\varphi_1, \ldots, \varphi_n \approx \psi?$$

A *probabilistic logic* (or *progic* for short) differs in two respects. First, the propositions have probabilities attached to them. Thus the premises have the form φ^X, where φ is a classical proposition and $X \subseteq [0,1]$ is a set of probabilities, and each premise is interpreted as 'the probability of φ lies in X'.[1] Second, the analogue of the classical question,

$$\varphi_1^{X_1}, \ldots, \varphi_n^{X_n} \approx \psi^Y?$$

is of little interest, because while there is often a natural conclusion ψ under consideration, there is rarely a natural probability set Y presented by the problem at hand since there are so many possible candidates for Y to choose from. Rather, the question of interest is the determination of Y itself:

$$\varphi_1^{X_1}, \ldots, \varphi_n^{X_n} \approx \psi^? \tag{1.1}$$

---

[1] This characterisation of probabilistic logic clearly covers what are called *external* progics in (Williamson, 2009b, §21)—the probabilities are metalinguistic, external to the propositions themselves. But it also covers *internal* progics, where the propositions involve probabilities (discussed in (Halpern, 2003), for example), and *mixed* progics, where there are probabilities both internal and external to the propositions.

R. Haenni et al., *Probabilistic Logics and Probabilistic Networks*, Synthese Library 350, DOI 10.1007/978-94-007-0008-6_1, © Springer Science+Business Media B.V. 2011

That is, what set Y of probabilities should attach to the conclusion sentence ψ, given the premises $\varphi_1^{X_1}, \ldots, \varphi_n^{X_n}$? This is a very general question, which will be referred to as the *Fundamental Question of Probabilistic Logic*, or simply as Schema (1.1).[2,3]

Part I of this book is devoted to showing that the fundamental question outlined above is indeed very general, providing a framework into which several common inferential procedures fit. Since the fundamental question of probabilistic logic differs from that of non-probabilistic logic, different techniques may be required to answer the two kinds of question. While proof techniques are often invoked to answer the questions posed in non-probabilistic logics, in Part II we show that probabilistic networks can help answer the fundamental question of probabilistic logic.

The programme of this book—namely that of showing how the fundamental question can (i) subsume a variety of inferential procedures and (ii) be answered using probabilistic networks—we call the *progicnet programme*. We view this as a logical programme. While it is possible to hold a narrow view of logic as being concerned primarily with proof—i.e., with the task of developing sound and complete axiomatisations—the progicnet programme is intended to fit with a broad view of logic as being concerned on the one hand with semantics—i.e., with specifying which inferences are in principle condoned by the logic—and on the other hand with calculi—i.e., with specifying one or more practical procedures for answering a given inferential question (e.g., truth-tables, semantic trees, proof systems, or, in our case, probabilistic networks).

1.2 The Potential of Probabilistic Logic

Due to the generality of Schema (1.1), many problem domains would benefit from an efficient means to answer its question—any problem domain whose structure has a natural logical representation and whose observations are uncertain in some respect. Here are some examples. In the philosophy of science we are concerned with the extent that a (logically complex) conclusion hypothesis is confirmed by a range of premise hypotheses and evidential statements which are themselves uncertain. In bioinformatics we are often interested in the probability that a complex molecule ψ is present, given the uncertain presence of molecules $\varphi_1, \ldots, \varphi_n$. In natural language processing we are interested in the probability that an utterance has semantic structure ψ given uncertain semantic structures of previous utterances and uncertain

---

[2] In asking what set of probabilities *should* attach to the conclusion, we are restricting our attention to logic rather than psychology. While the question of how humans go about ascribing probabilities to conclusions in practice is a very interesting question, it is not one that we broach in this book.

[3] In our notation, the probabilities are attached to the propositions $\varphi_1, \ldots, \varphi_n, \psi$, *not* to the entailment relation. However, in the literature one sometimes sees expressions of the form $\varphi_1^{X_1}, \ldots, \varphi_n^{X_n} \approx^Y \psi$ (Williamson, 2002, §2.2–2.3). Our choice of notation is largely a question of convenience: in our notation the premises and conclusion turn out to be the same sort of thing, namely propositions with attached probabilities, and there is a single entailment relation rather than an uncountable infinity of entailment relations; but of course from a formal point of view the two kinds of expression can be used interchangeably.

contextual factors. In robotics we are interested in finding the sequence of actions of a robot that is most likely to achieve a goal given the uncertain structure of the robot's surroundings. In expert systems we are interested in the probability to attach to some prediction or diagnosis given statistical knowledge about past cases. The list goes on.

Unfortunately, this potential of probabilistic logics has not yet been exploited. There are a number of reasons for this. First, current probabilistic logics are a disparate bunch—it is hard to glean commonalities to see how they fit into a general framework, and hard to see how a solution to the general problem of probabilistic logic would specialise to each individual logic (Williamson, 2002, 2009b, §21). Second, probabilistic logics are often hard to understand: while probabilistic reasoning is well understood and so is logical reasoning, when these two components interact in formalisms that combine them, their complexities compound and a great deal of theoretical work is required to determine their properties. Third, probabilistic logics are often thwarted by their computational complexity. While they may integrate probability and logic successfully, it may be very difficult to determine an answer to a question such as that of Schema (1.1). Sometimes this is because a probabilistic logic seeks more generality than is required for applications; but often it is no fault of the logic—probabilistic and logical reasoning are both computationally infeasible in the worst case, and their combination is no more tractable.

1.3 Overview of the Book

In this book we hope to address some of these difficulties. In Part I we show how a range of alternative inferential procedures fit into a general framework for probabilistic logic. We will cover the standard probabilistic semantics for probabilistic logic in §2, in §3 the support-possibility approach of the probabilistic argumentation framework, evidential probability in §4, inference involving statistical hypotheses in §5 and §6, and objective Bayesian epistemology in §7. The background to each procedure will be discussed in §X.1, where X ranges from 2 to 7; note that §2.1 contains prerequisites for the other sections and should not be skipped on a first reading. In §X.2 we show how a key question of each inferential procedure can be viewed as a question of the form of Schema (1.1). In §X.3 we will show the converse, namely that each inferential procedure can be viewed as providing semantics for the entailment relation $\approx$ found in this schema.

We permit a generic notion of entailment $\approx$ which is weaker than that of classical logic. Generally, the entailment $\varphi_1^{X_1}, \ldots, \varphi_n^{X_n} \approx \psi^Y$ holds iff all models of the left-hand side satisfy the right-hand side, where suitable notions of *model* and *satisfy* are filled in by the semantics in question. We say that a semantics for the entailment relation yields a *probabilistic logic* if (i) models are probability functions (satisfying certain conditions that are specified by the semantics) and (ii) probability function P satisfies ψ^Y iff $P(\psi) \in Y$. In this book we distinguish between *non-monotonic* and *monotonic* entailment relations. Monotonicity holds where $\varphi_1^{X_1}, \ldots, \varphi_n^{X_n} \approx \psi^Y$

implies $\varphi_1^{X_1}, \ldots, \varphi_n^{X_n}, \ldots, \varphi_m^{X_m} \mathrel{|\approx} \psi^Y$ for $m \geq n$. Entailment under the standard se-mantics is monotonic, for example, whereas (first-order) evidential probability and objective Bayesian epistemology are non-monotonic. We can call an entailment re-lation $\mathrel{|\approx}$ *decomposable* if $\varphi_1^{X_1}, \ldots, \varphi_n^{X_n} \mathrel{|\approx} \psi^Y$ implies that each model of each of the premises individually is a model of the conclusion: for all interpretations I (as defined by the semantics in question), if $I \mathrel{|\approx} \varphi_1^{X_1}, \ldots, I \mathrel{|\approx} \varphi_n^{X_n}$ then $I \mathrel{|\approx} \psi^Y$. A de-composable entailment relation is monotonic, but the reverse need not be the case. The standard semantics provides an example of a decomposable entailment relation.

Roughly speaking, the inferential procedures considered in this book provide the following differing semantics for the entailment relation. Under the standard seman-tics, a model is simply a probability function defined over the logical language of the propositions in the premises and conclusion, and $\varphi_1^{X_1}, \ldots, \varphi_n^{X_n} \mathrel{|\approx} \psi^Y$ iff each probability function that satisfies the left-hand side also satisfies the right-hand side, i.e., iff each probability function P for which $P(\varphi_1) \in X_1, \ldots, P(\varphi_n) \in X_n$ yields $P(\psi) \in Y$. In the probabilistic argumentation framework, one option for the en-tailment to hold is if Y contains all the probabilities of the worlds for which the left-hand side *forces* ψ to be true. According to *second-order* evidential probability, where the φ_i are statistical statements and logical relationships between classes and ψ is the assignment of *first-order* evidential probability on those premises, the en-tailment holds if whenever the risk-level of each ϕ_i is contained in X_i, the risk-level of ψ is contained in Y. According to fiducial probability, the premises can either be spelled out in terms of functional models and data, where the conclusion concerns a bandwidth of probability and the entailment holds if the data and model warrant the bandwidth, or in terms of an assignment of first-order evidential probability. According to Bayesian statistical inference the premises contain information about prior probabilities and likelihoods which constitute a statistical model, the conclu-sion denotes posterior probabilities, and the entailment holds if for every probability function subsumed by the statistical model of the premises, the conclusion follows by Bayes' theorem. According to objective Bayesian epistemology, the entailment holds if some probability function P gives $P(\psi) \in Y$, from those functions that sat-isfy the constraints imposed by the premises and are otherwise maximally equivocal.

The various inferential procedures covered in this book provide *different* and largely incompatible semantics for probabilistic logic, nevertheless they have some things in common. First, in each case the premises on the left-hand side of Schema (1.1) are viewed as *evidence*, while the proposition ψ of the conclusion is a hypothesis of interest. Second, each account admits of a formal connection to the mathematical concept of probability. We use the term *probability* exclusively in this mathematical sense of a measure that satisfies the usual Kolmogorov axioms for probability. While from a conceptual point of view several accounts may distance themselves from this standard notion of probability, they retain a formal connection to probability, and it is this connection that can be exploited to provide a syntactic procedure for determining an answer to the question of Schema (1.1).[4] This task

---

[4] We should emphasize that we do not seek to revise the conceptual foundations of each approach—our point is that despite their disparate philosophies, these approaches have a lot in common from a formal point of view, and that these formal commonalities can be harnessed for inference.

of determining an answer to Schema (1.1) is the goal of Part II. The syntactic and algorithmic nature of this task are points in common with the notion of proof in classical logic, but as we saw at the start of this chapter, the question being asked by Schema (1.1) is slightly different to that being asked of classical logic.

We will pay particular attention to the case in which the sets of probabilities $X_1, \ldots, X_n, Y$ are all taken to be *convex*, i.e., *sub-intervals* of $[0, 1]$. The advantage of this framework is that it is general enough to cover many interesting approaches to combining probability and logic, while being narrow enough to take a serious stab at the computational concerns. It is not as general as it might be: the sets $X_1, \ldots, X_n, Y$ could be taken to be arbitrary sets of probabilities, or sets of gambles bounded by lower and upper previsions (de Cooman and Miranda, 2007, §2.1–2.2), but most approaches only require convex sets of probabilities.[5] Moreover, by focussing on convex sets, we can apply the machinery of probabilistic networks to address the computational challenge. In Part II we shall show how credal and Bayesian networks can be applied to more efficiently answer questions of the form of Schema (1.1). §8 presents common machinery for using a probabilistic network to answer the fundamental question. The algorithm for constructing the network itself depends on the chosen semantics and is discussed in subsequent sections of Part II.

Finally, we want to emphasize that the probabilistic logic that is to be presented in this book is not merely an exercise in logical versatility. As illustrated in Romeijn et al. (2009), it can be used to solve real problems in statistics, especially when the statistical methods are required to incorporate background knowledge that is expressed by logical constraints. More generally, a unified probabilistic logic may contribute to a more efficient use of probabilistic methods in science, computation, and statistics: it brings logical constraints within the reach of these probabilistic methods, and by means of credal networks it offers computationally feasible methods to mesh with these constraints. We hope that this will contribute to the use of probabilistic methods in much in the way that the notion of causality and its relation to Bayesian networks have contributed to scientific, computational, and methodological advances.

1.4 Philosophical and Historical Background

The approaches under consideration here take different stances as to the interpretation of probability. The standard semantics leaves open the question of the nature of probabilities—any interpretation can be invoked. Probabilities in the probabilistic argumentation framework are also not tied to any particular interpretation, but then the degree of support of a proposition is interpreted as the probability of the scenarios in which the evidence forces the truth of the proposition. Evidential probability and classical statistical inference are based on the frequency interpretation of probability. Bayesian statistical inference is developed around the use of Bayes'

---

[5] One might even generalise further by taking $X_1, \ldots, X_n, Y$ to be arbitrary representations of uncertainty, but if we forfeit a connection with probability, we leave the realm of probabilistic logic.

rule, which requires prior probabilities; these prior probabilities are often given a Bayesian—i.e., degree-of-belief—interpretation, but the formal apparatus in fact permits other interpretations. On the other hand objective Bayesianism fully commits to probabilities as degrees of belief, and, following Bayes, these probabilities are highly constrained by the extent and limits of the available evidence.

We should emphasise that the probabilistic logic to be presented here is certainly not the first one around. In fact the history of the notion of probability is intimately connected to probabilistic logics, or systems of reasoning with probability. According to Galavotti (2005), the first explicit versions of a probabilistic logic can be found in nineteenth century England, in the works of De Morgan, Boole, and Jevons. Generally speaking, they perceived probability as a measure of partial belief, and they thought of probability theory as providing an objective and normative guide to forming conclusions on the basis of partial beliefs. Logicist probabilists of the early twentieth century, such as Keynes, Koopman and Ramsey, developed their ideas from these two starting points. While Keynes took probability theory as describing the rules of partial entailment, and thus as the degree to which evidence objectively supports some conclusion, Ramsey took it as providing rules for maintaining coherent partial beliefs, thus leaving room for differing opinions and surrendering some objectivity to the subjectivity of an individual's beliefs.

The logical interpretation of probability, as advanced by Keynes and Koopman, was rarely embraced by scientists in the early twentieth century, although Harold Jeffreys can be viewed as an important exception. But the approach was eventually picked up by philosophers, most notably in the probabilistic logics proposed by Carnap (1950, 1952, 1980) and his followers. Carnap's systems focus primarily on a logical relationship between an hypothesis statement and an evidence statement, and are one approach to formalizing Keynes's idea that probability is an objective measure of partial entailment. The subjective view of Ramsey, on the other hand, has become progressively more popular in the latter half of the twentieth century. Developed independently by Savage and de Finetti, this view of probability has gained popularity among probabilistic logicians with an interest in inductive reasoning, with Howson (2001) as a strong representative, and among decision theorists such as Jeffrey (1965).

It soon became apparent that there were several kinds of question raised by probability—e.g., whether ignorance is distinguishable from risk, whether assessments of probability are distinguishable from decision, and what role consistency and independence play in probability assignment. The Dempster-Shafer theory (Shafer, 1976) and the theory of Imprecise Probability (Walley, 1991) are two important theories embracing the first distinction, which raises ramifications for the remaining two, all of which were the subject of work on interval-valued probability in the post-war era by Tarski, C.A. Smith, Gaifman, Levi, Kyburg, Raiffa, and Arrow.

Recent work in artificial intelligence has contributed to our theoretical understanding of probability logic, particularly with the work of Fagin, Halpern, Bacchus, Grove, Koller, and Hailperin (Bacchus et al., 1993; Fagin and Halpern, 1991, 1994; Halpern and Fagin, 1992; Hailperin, 1996), (Kyburg, 1987; Kyburg and Pittarelli,

1996; Kyburg and Teng, 2001), Lehman and Magidor (1990), Pearl (1988, 1990b), Pollock (1993), Nilsson (1986), as well as our practical understanding of its application to learning from data (Neapolitan, 1990, 2003), causal reasoning (Pearl, 1988; Spirtes et al., 1993), multi-agent systems (Fagin et al., 2003), robotics (Thrun et al., 2005), logic programming (Kersting and Raedt, 2007), among other fields.

1.5 Notation and Formal Setting

In this book we will primarily be focussing on sets of variables, propositional languages, and simple predicate languages. Logical languages will be denoted by the letter $\mathscr{L}$. We will represent variables by capital letters near the beginning of the alphabet, A, B, C, etc. A *propositional variable* is a variable A that takes one of two possible values, *true* or *false*. The notation a or a^1 will be used to denote the assignment $A = true$, while $\bar{a}$ or a^0 signifies $A = false$. Given propositional variables $A_1, \ldots, A_n$, a propositional language contains sentences built in the usual way from the assignments $a_1, \ldots, a_n$ and the logical connectives $\neg$, $\wedge$, $\vee$, $\rightarrow$, and $\leftrightarrow$. An *elementary outcome* ω is an assignment $a_1^{e_1} \cdots a_n^{e_n}$ where $e_1, \ldots, e_n \in \{0, 1\}$. An *atomic state* α is a sentence that denotes an elementary outcome: α is a conjunction $\pm a_1 \wedge \cdots \wedge \pm a_n$ where $\pm a_i$ is a_i (respectively $\neg a_i$) if $e_i = 1$ (respectively $e_i = 0$) in the elementary outcome. Given $e = (e_1, \ldots, e_n)$, we let α^e denote the atomic state describing the elementary outcome $a_1^{e_1} \cdots a_n^{e_n}$. Thus superscripts are used to describe particular assignments of values to variables.

Predicates will be represented by U, V, W, constants by $t, t_1, t_2, \ldots$, and logical variables by x, y, z. Expressions of the form $U(t)$ and $V(x)$ determine (single-case and, respectively, repeatably-instantiatable) propositional variables; Ut and Vx will be used to denote the positive assignments $U(t) = true$ and $V(x) = true$. Finitely many such atomic expressions yield a propositional language. Sentences of a logical language will be denoted by Greek letters φ, ψ, etc., and capital Greek letters—e.g., Γ, Δ, Θ, Φ—will be used for sets of sentences. Again, the letter α will be reserved for an *atomic state* or *state description*, which is a conjunction of atomic literals, e.g., $Ut_1 \wedge \neg Vt_2 \wedge \neg Wt_3$, where each predicate in the language (or in a given finite sublanguage) features in α. Entailment relations are denoted as shown in Table 1.1.

Table 1.1: Entailment Relations

Symbol	Entailment Relation
$\mathrel{\vert\!\approx}$	generic entailment
$\mathrel{\vert\!\sim}$	non-monotonic entailment
$\mathrel{\vert\!\approx}$	monotonic entailment
$\models$	decomposable, monotonic entailment

We shall use P, Q, R, S to denote probability functions, $\mathbb{P}$ for a set of probability functions, and $\mathbb{K}$ for a credal set, i.e., a closed convex set of probability functions. X, Y, Z will be sets of probabilities and ζ, η, θ parameters in a probabilistic model. E, F, G, H will be used for subsets and $\omega, \omega_1, \omega_2, \ldots$ for elements of the outcome space Ω. Algebras of such subsets, finally, are denoted by $\mathscr{E}, \mathscr{F}$.

Chapter 2
Standard Probabilistic Semantics

What we call the *standard probabilistic semantics* (or *standard semantics* for short) is the most basic semantics for probabilistic logic. According to the standard semantics, an entailment relation

$$\varphi_1^{X_1}, \ldots, \varphi_n^{X_n} \approx \psi^Y$$

holds if all probability functions that satisfy the constraints imposed by the left-hand side also satisfy the right. The standard semantics serves as a starting point for comparing the different interpretations of Schema (1.1). It is very much in the same vein as the probability logic proposed in (Hailperin, 1996).

In §2.1 we introduce probability functions, interval-valued probabilities and imprecise probabilities. §2.2 shows that the key question facing the standard semantics is naturally represented as a question of the form of Schema (1.1). §2.3 shows the converse, namely that a question of the form of Schema (1.1) can be naturally interpreted by appealing to the standard semantics.

2.1 Background

The standard semantics comprises the traditional logical tenet that inference rules must be truth-preserving, and the further tenet that for probabilistic logic the formal models are probability measures and thus comply to the axioms of Kolmogorov (1950). Neither tenet is common to all the perspectives in this book, but it is instructive to note how each account diverges from the standard semantics.

R. Haenni et al., *Probabilistic Logics and Probabilistic Networks*, Synthese Library 350, 11
DOI 10.1007/978-94-007-0008-6_2, © Springer Science+Business Media B.V. 2011

2.1.1 Kolmogorov Probabilities

The standard probabilistic semantics for a single measure on a propositional language is provided in terms of a probability structure, which we shall define shortly. A probability structure is based upon a specified probability space.

Definition 1 (Probability Space). A *probability space* is a tuple $(\Omega, \mathscr{F}, P)$, where Ω is a sample space of elementary events, $\mathscr{F}$ is a σ-algebra of subsets of Ω, and $P : \mathscr{F} \rightarrow [0,1]$ is a probability measure satisfying the Kolmogorov axioms:

(P1) $P(E) \geq 0$, for all $E \in \mathscr{F}$;
(P2) $P(\Omega) = 1$;
(P3) $P(E_1 \cup E_2 \cup \cdots) = \sum_i P(E_i)$, for any countable sequence $E_1, E_2, \ldots$ of pairwise disjoint events $E_i \in \mathscr{F}$.

Note that when Ω is finite and E and F are disjoint members of $\mathscr{F}$, a special case of P3 is

(P3*) $P(E \cup F) = P(E) + P(F)$,

from which, along with P1, a useful general additivity property may be derived, namely

(P3′) $P(E \cup F) = P(E) + P(F) - P(E \cap F)$.

Definition 2 (Probability Structure). A *probability structure* is a quadruple $M = (\Omega, \mathscr{F}, P, I)$, where $(\Omega, \mathscr{F}, P)$ is a probability space and I is an interpretation function associating each elementary event $\omega \in \Omega$ with a truth assignment on the propositional variables Φ in a language $\mathscr{L}$ such that $I(\omega, A) \in \{true, false\}$ for each $\omega \in \Omega$ and for every $A, B, C, \ldots \in \Phi$.

Since P is defined on events rather than sentences, we need to link events within a probability structure M to formulas in Φ. If we associate $[\![\varphi]\!]_M$ with the set of elementary events within (finite) Ω in M where φ is true, then the following proposition makes explicit the relationship between formulas and events.

Proposition 1. *For arbitrary propositional formulas φ and ψ,*

1. $[\![\varphi \wedge \psi]\!]_M = [\![\varphi]\!]_M \cap [\![\psi]\!]_M$,
2. $[\![\varphi \vee \psi]\!]_M = [\![\varphi]\!]_M \cup [\![\psi]\!]_M$,
3. $[\![\neg\varphi]\!]_M = \Omega \setminus [\![\varphi]\!]_M$.

Under this interpretation, the assignments of probability to sets in the algebra are effectively assignments of probability to expressions in $\mathscr{L}$. Hence,

$$P(\varphi) \Leftrightarrow P([\![\varphi]\!]_M).$$

For present purposes we may compress notation by using capital letters to denote both propositions within a language $\mathscr{L}$ and the corresponding events within $\mathscr{F}$ of some structure M, and we may also omit the subscript M when the context is clear.

Ronald Fagin, Joseph Halpern, and Nimrod Megiddo (Fagin et al., 1990) provide a proof theory for the standard semantics on a propositional language. Deciding satisfiability is NP-complete. There are obstacles to providing a proof theory for probability logics on more expressive languages, however. Halpern (1990) discusses a first-order probability logic allowing φ to represent probability statements, but these systems are highly undecidable. Indeed, the validity problem for first-order probability logic with a single binary predicate is not even decidable relative to the full theory of real analysis. The reason is that standard Kolmogorov probability is a higher-order function on sets, so a language that is expressive enough to afford probabilistic reasoning about probability statements will extend beyond the complexity of first-order reasoning about real numbers and natural numbers.

2.1.2 Interval-Valued Probabilities

One may feel that the above is really all there is to probabilistic logic. That is, one may think that the axioms of probability theory and the interpretation of the algebra as a language already provide a complete logic. Bayesian probabilistic logic, as put forward in Ramsey (1926) and De Finetti (1937), and explicitly advocated by Howson (2001, 2003), Morgan (2000), and Halpern (2003), is exactly this. In this logic we interpret the probability assignments as a kind of partial truth valuation, or as a degree of belief measured by betting quotients. The axioms may then be taken as the sole consistency constraints on these valuations, and thus as inference rules in the language. Apart from these constraints we are allowed to choose the probability assignments over the algebra freely, as long as they have sharp values.

As brought out clearly in Hailperin (1996), the latter requirement is by no means a necessity for this type of probabilistic logic. Let us examine some cases in which logical formulas cannot be assigned sharp probabilities. The first case occurs within the standard semantics itself. From P3' we may derive constraints for events E and F in $\mathscr{F}$ even when we do not know the value of $P(E \cap F)$. For example, if $P(E) = 0.6$ and $P(F) = 0.7$, and this is all that is known about E and F, then we may derive that $0.7 \leq P(E \vee F) \leq 1$, and that $0.3 \leq P(E \wedge F) \leq 0.6$. This constraint is generalized by the following proposition (Wheeler, 2006).

Proposition 2. *If $P(E)$ and $P(F)$ are defined in M, then:*

1. $P(E \cap F)$ lies within the interval

$$[\max(0, (P(E) + P(F)) - 1), \min(P(E), P(F))], \text{ and}$$

2. $P(E \cup F)$ lies within the interval

$$[\max(P(E), P(F)), \min(P(E) + P(F), 1)].$$

So, the standard semantics allows for interval-valued conclusions on the basis of sharp-valued assignments in the premises.

It thus seems that interval-valued probability assignments follow rather naturally from the standard semantics using sharp assignments. We might then ask how to extend the use of interval-valued assignments to premises. The strategy for dealing with this case is exactly the same: the premises can still be seen as restrictions on a set of probability assignments. But there are several tactical options to consider before specifying how to generalize the way the standard semantics handles the general question behind Schema (1.1).

One way of defining such interval-valued probability assignments is by means of inner and outer measures (Halmos, 1950; Walley, 1991; Halpern, 2003; Wheeler, 2006). Suppose we do not have a sharp probability value for an event F because $F \notin \mathscr{F}$ within our probability structure, but F is logically related to events in $\mathscr{F}$. For instance, suppose we know that F contains E and that F is contained within G, and that both E and G are within $\mathscr{F}$. When there is no measurable event contained in F that dominates E, then E is a *kernel event* for F. When every measurable event containing F dominates G, then G is a *covering event* for F. The measures of F's kernel and cover then yield non-trivial bounds on F with respect to M, since otherwise $P(F)$ would be undefined.

We express this idea in terms of inner and outer measures. If measure P is defined on $\mathscr{F}$ of M and E' is not in $\mathscr{F}$, then $P(E')$ is not defined since E' isn't in the domain of P. However E' may be an element of an algebra $\mathscr{F}'$ such that $\mathscr{F}$ is a subalgebra of $\mathscr{F}'$. We may then extend the measure P to the set E' by defining inner and outer measures to represent our uncertainty with respect to the precise measure of E'.

Definition 3 (Inner and Outer Measure). Let $\mathscr{F}'$ be a subalgebra of an algebra $\mathscr{F}$, $P : \mathscr{F} \to [0,1]$ a probability measure defined on the space $(\Omega, \mathscr{F}, P)$, and E an arbitrary set in $\mathscr{F}'$. Then define the *inner measure $\underline{P}$ induced by P* and the *outer measure $\overline{P}$ induced by P* as:

$$\underline{P}(E) = \sup\{P(F) : F \subseteq E, F \in \mathscr{F}\} \text{ (inner measure of } E);$$
$$\overline{P}(E) = \inf\{P(F) : F \supseteq E, F \in \mathscr{F}\} \text{ (outer measure of } E).$$

We now observe some properties of inner and outer measures:

(P4) $\underline{P}(E \cup F) \geq \underline{P}(E) + \underline{P}(F)$, when E and F are disjoint (superadditivity);
(P5) $\overline{P}(E \cup F) \leq \overline{P}(E) + \overline{P}(F)$, when E and F are disjoint (subadditivity);
(P6) $\underline{P}(E) = 1 - \overline{P}(\overline{E})$;
(P7) $\underline{P}(E) = \overline{P}(E) = P(E)$, if $E \in \mathscr{F}$.

Properties P4 and P5 follow from P2 and P3. Note that when Ω is finite, P4 and P5 follow from P2 and P3*. P6 makes explicit the relationship between inner and outer measures. By P3, for each set E, there are measurable sets $F, G \in \mathscr{F}$ such that $F \subseteq E \subseteq G$ and $\underline{P}(E) = P(F)$ and $\overline{P}(E) = P(G)$. Note then the limit cases: if there are no measurable sets containing E other than the entire space Ω, then $\overline{P} = 1$; if there are no non-empty measurable sets contained in E, then $\underline{P}(E) = 0$. Thus, P6 allows us to represent the situation in which we are entirely ignorant of event

E. P7 makes explicit that inner and outer measures strictly extend *P*: if an event *E* is measurable, then the inner (outer) measure of *E* is $P(E)$. Thus, P6 represents the case when we have sharp probabilities for *E*. Finally by P2 and P3, we may generalize P4 to

(P4′) $\underline{P}(E \cup F) \geq \underline{P}(E) + \underline{P}(F) - \underline{P}(E \cap F)$ (generalized superadditivity).

A positive function satisfying P2 and P4′ is called a *2-monotone Choquet capacity*, which may be generalized to an *n-monotone Choquet capacity* when P4′ is replaced by

$$(P4^*)\ \underline{P}\left(E = \bigcup_{i=1}^{n} E_i\right) \geq \sum_{i=1}^{n} \sum_{\{F \subseteq \{E_1,\dots,E_n\} : |F| = i\}} (-1)^{i+1} \underline{P}\left(\bigcap_{i+1} E_{i+1}\right).$$

P4* says that the inner-measure of the union of *n* events is greater or equal to the sum of adding all marginal inner measures $\underline{P}(E_1) + \cdots + \underline{P}(E_n)$, subtracting all pairs of intersections in *E*, adding all 3-member intersections, and so on, alternating through to *n*. The switch between addition of odd intersections and subtraction of even intersections is handled by the $(-1)^{i+1}$ term. Note that Proposition 2 records properties for 1-monotone capacities. A 1-monotone probability logic is studied in (Wheeler, 2006). Finally, a Dempster-Shafer belief function is an ∞-monotone capacity.

2.1.3 Imprecise Probabilities

We now turn to the relationship between inner-measures and sets of probabilities.

Theorem 1 (Horn and Tarski, 1948). *Suppose P is a measure on a (finitely additive) probability structure $\mathscr{M}$ such that $\mathscr{F} \subseteq \mathscr{F}'$. Define $\mathbb{P}$ as the set of all extensions P' of P to $\mathscr{F}'$. Then for all $E \in \mathscr{F}'$:*

1. $\underline{P}(E) = \underline{\mathbb{P}}(E) = \inf\{P'(E) : P' \in \mathbb{P}\}$, and
2. $\overline{P}(E) = \overline{\mathbb{P}}(E) = \sup\{P'(E) : P' \in \mathbb{P}\}$.

The Horn-Tarski result links the inner-measure of an event *E* to the lower probability $\underline{\mathbb{P}}(E)$ for a *particular* set of probability measures, namely those which extend $\mathscr{F}'$ to events in $\mathscr{F}$. (A lower probability $\underline{P}$ satisfies (P1), (P2), (P4), and (P6).) However, the lower probability of an arbitrary set of measures does not necessarily satisfy (P4*). So, lower probability in general is not equivalent to the lower envelope of the set $\mathbb{P}$ extending $\mathscr{F}'$ to $\mathscr{F}$.

To go from a lower probability to the lower envelope of a set $\mathbb{P}$ of distributions, observe that every 2-monotone lower probability is a lower envelope. That means that this set of probabilities that form the lower envelope is the set of probabilities that dominate the 2-monotone lower probability, where a probability distribution $P(E)$ dominates a lower probability $\underline{P}(E)$ if $P(E) \geq \underline{P}(E)$. Since the set $\mathbb{P}^*$ of measures *P* that dominates $\underline{P}(E)$ is a (possibly empty) closed, convex polyhedron within the space of possible measures (Walley, 1991), we might think that dominated lower probability in general is sufficient to identify the convex hull. But it isn't. The problem is that $\mathbb{P}$ may not satisfy (P4*) and one can construct dominated, 1-monotone lower probabilities that fail to be a lower envelope. But every 2-monotone lower probability is a lower envelope (Walley, 1991, §3.3.3). So, if we find the set of probability distributions that dominate the 2-monotone lower probability of a set $\mathbb{P}$ we will have found the lower envelope of $\mathbb{P}$. Another way to put this is that although a set $\mathbb{P}^* \subsetneq \mathbb{P}$ of dominated measures is a convex set, the convex hull of $\mathbb{P}^*$ is not necessarily the convex hull of $\mathbb{P}$. We need the additional monotonicity condition. Before pressing on, first a short digression on convex sets. For convexity is not only important for linking lower probability to a set of distributions, but the property is also important for the network structures we consider in Part II.

2.1.4 Convexity

Generally, a set X is called convex if and only if X is closed under all binary operations $b_{(\lambda,1-\lambda)}$ for $\lambda \in [0,1]$. To understand this notion, picture two points, x, y, that are members of a real line segment X defined on a linear space Z. The *straight line segment* from x to y is the set $\{\lambda x + (1-\lambda)y : \lambda \in [0,1]\}$. This equation specifies the convex hull of the set $\{x, y\}$, and the endpoints x and y are its extremal points. A set $X \subseteq Z$ is convex if and only if X contains the straight line segment connecting each pair of its members.

A *credal set* $\mathbb{K}$ is a convex set of probability functions, i.e. $P_1, P_2 \in \mathbb{K}$ implies $\lambda P_1 + (1-\lambda)P_2 \in \mathbb{K}$ for all $\lambda \in [0,1]$. To illustrate, supposes that an agent's belief state is represented by $\mathbb{K}$. Then for any P_1 and P_2 contained in $\mathbb{K}$ and any $0 \le \lambda \le 1$, $\mathbb{K}$ also contains P_3 such that

$$P_3 = \lambda P_1(E) + (1-\lambda)P_2(E),$$

for any E in the algebra. This view is in contrast to strict Bayesianism, which is committed to a single probability function and numerically determinate degrees of belief. Isaac Levi, who pioneered *convex Bayesianism*, has shown that the set formed by updating elements of $\mathbb{K}$ by conditionalization is also convex (Levi, 1980). Thus, convex Bayesianism is a generalization of strict Bayesianism, since numerically determinate degrees of belief are the special case where $\mathbb{K}$ is a singleton. Further, Henry Kyburg (1987) has shown that any Dempster-Shafer belief function corresponds to a convex set of probability functions, but not conversely. In short, convex Bayesianism includes both strict Bayesianism and Dempster-Shafer belief functions (Shafer, 1976) as special cases.

There is also a relationship between a convex set of probability functions and lower probability.

Theorem 2 (Walley 1991). *If $\mathbb{K}$ is a convex set of probability functions, then*

1. $\underline{P}(E) = \underline{\mathbb{K}}(E) = \inf\{P(E) : P \in \mathbb{K}\}$; and

2. $\overline{P}(E) = \overline{\mathbb{K}}(E) = \sup\{P(E) : P \in \mathbb{K}\}$.

This correspondence is proved by Walley (1991, §3.3.3).

All the frameworks that we consider in this book will exploit convex sets of probabilities in one fashion or another, but care should be exercised establishing the conditions under which convexity holds. This is particularly important when considering evidential probability in §4 and statistical reasoning in §5.

For example, Kyburg and Pittarelli (1996) describe a scenario in which you are told that a 6-sided die is fair for outcomes '3', '4', '5', and '6', but is *either* biased $1/12$ against '1' and $3/12$ in favor of '2' *or* biased $3/12$ in favor of '1' and $1/12$ against '2'. The possible baises for outcomes $(1,2,3,4,5,6)$ is the set

$$\{(1/12, 3/12, 1/6, 1/6, 1/6, 1/6), (3/12, 1/12, 1/6, 1/6, 1/6, 1/6)\},$$

which is not convex. If we take the set to represent possible frequencies, then by convex Bayesianism any betting odds on outcome '1' between 1:11 to 3:9 is acceptable. While any single bet in this range may be reasonable, we know that in the long run only one of the two extremal points, $1/12$ or $3/12$, will represent fair odds for a sequence of bets and that long-term betting at any other fixed odds will ensure sure loss. Of course one might expect betting odds to tend to one or other extremal point in the light of observed outcomes of rolls of the die, in which case sure loss is avoided. But, this essentially ignores the information provided from the start that the die is biased one way or the other.

Teddy Seidenfeld et al. (2003) have shown the non-equivalence of 3-way versus pair-wise comparisons of admissible options in decision problems, and they have presented an axiomatized theory of coherent choice functions without convexity to avoid these counterexamples (Seidenfeld et al., 2007).

Finally, observe that the convexity of a set by definition depends on the coordinates that determine which lines are taken as straight in a certain space. In the case of credal sets, the convexity of a set thus depends on the coordinates of the space of probability assignments. In §8.1.2 we will illustrate this further.

Although the methods we develop in this book focus on convex sets of probability distributions, it is far from clear that a set-based probability logic must be a *convex* set probability logic (Cozman, forthcoming). Joe Halpern (2003, §2.3) discusses the relationship between lower probability and Choquet capacities without assuming convexity, and Seidenfeld et al.'s axiomatization without convexity (Seidenfeld et al., 2007) provide important insights into working with sets of distributions. So, while we will exploit the properties of convex sets of distributions we are well aware of their limitations and (with optimism) leave for future research the question of adapting our methods to non-convex methods.

2.2 Representation

It is clear that the key question of what Y to attach to ψ is a question of the form of Schema (1.1). For all P defined on the propositional language $\mathscr{L}$, if $P(\varphi_1) \in X_1, \ldots,$ $P(\varphi_n) \in X_n$ entails $P(\psi) \in Y$, then

$$\varphi_1^{X_1}, \ldots, \varphi_n^{X_n} \models \psi^Y.$$

The entailment relation employed in the standard semantics is classical consequence. In terms of our classification of entailment relations, the entailment under the standard semantics is monotonic and decomposable. The only difference between the standard probabilistic semantics and the standard semantics for propositional logic is that the models for the probabilistic logic are full probability assignments over the language rather than truth assignments. An inference in a framework of probabilistic logic is valid iff the set of all models, i.e. probability assignments, satisfying the premises is included in the set of models, i.e. probability assignments, satisfying the conclusion, as depicted in Figure 2.1.

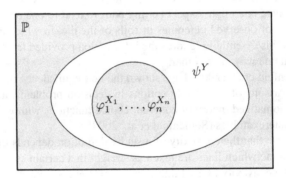

Fig. 2.1: Validity in the standard semantics comes down to the inclusion of the set of probability functions satisfying the premises, $\varphi_1^{X_1}, \ldots, \varphi_n^{X_n}$, in the set of assignments satusfying the conclusion, ψ^Y. The rectangular space of probability assignments $\mathbb{P}$ includes all probability assignments over the language $\mathscr{L}$.

Normally premises are of the form a^X, presenting a direct restriction of the probability for a to X, that is, $P(a) \in X \subset [0,1]$, where X might also be a sharp probability value $P(a) = x$. In a few cases, however, the restrictions imposed by the premises can take alternative forms. An example is the premise $(a|b)^x$, meaning that $P(a|b) = x$. Such premises cannot be interpreted directly as the measures of specific sets in the associated semantics. Instead they present restrictions to the ratio of the measures of two sets. By definition we have $P(a|b) = P(a \wedge b)/P(b)$, so we may say that $(a|b)^x$ is a shorthand form of two normal premises which together entail the restriction, namely

$$(a|b)^x \;\Leftrightarrow\; \forall y \in (0,1] : b^y, (a \wedge b)^{xy}. \tag{2.1}$$

Restrictions to the probability assignment P that can be spelled out in terms of combinations of inter-related normal premises a^x we will call *composite*. In principle, the standard semantics allows for any premise that can be understood as a composite restriction. See §3 and §6 for examples.

Some special attention must be devoted to premises to do with independence relations between propositional variables. Examples are $A \perp B$, meaning that $P(A,B) = P(A)P(B)$, or $A \perp B|C$, which means that $P(A,B|C) = P(A|C)P(B|C)$. These more complex probabilistic relations can also be incorporated into the premises, for example by

$$\exists x, y \in [0,1] : a^x, b^y, (a \wedge b)^{xy}, (a \wedge \bar{b})^{x(1-y)}, (\bar{a} \wedge b)^{(1-x)y} \qquad (2.2)$$

for $P(A,B) = P(A)P(B)$. Equation (2.2) can be combined with Equation (2.1) to provide the probabilistic restriction associated with $A \perp B|C$. Examples of such premises can be found in §8 and §13.

2.3 Interpretation

We may also use the standard probabilistic semantics to provide an interpretation of

$$\varphi_1^{X_1}, \ldots, \varphi_n^{X_n} \approx \psi^Y.$$

In this interpretation, the premises provide constraints on a probability assignment, and the conclusion is a constraint that is guaranteed to hold if all the constraints of the premises hold. Clearly $\varphi_1^{X_1}, \ldots, \varphi_n^{X_n} \approx \psi^{[0,1]}$, so the problem isn't to find *some* Y that $\varphi_1^{X_1}, \ldots, \varphi_n^{X_n} \approx \psi^Y$ but to find the *smallest* such Y. Since any superset of this minimal Y can also be attached to the conclusion sentence ψ, if one finds the minimal Y then one determines all Y for which the entailment relation holds.

From our observations, the standard semantics answers the fundamental question posed by Schema (1.1) by finding the lower and upper envelope of the convex set $\mathbb{K}$ of probability functions which satisfy all premises on the left-hand side of Schema (1.1). So the standard semantics deals with interval-valued assignments to premises in exactly the way we should expect. This is illustrated in the following two examples.

Example 1. If we have $\varphi_1^{0.2}$ and $\varphi_2^{[0.3,0.4]}$ (without further specifying φ_1 and φ_2), then the models include all probability measures P for which $P(\varphi_1) = 0.2$ and $0.3 \leq P(\varphi_2) \leq 0.4$. Now imagine that we ask $(\varphi_1 \vee \varphi_2)^?$. What minimal interval Y can we attach to $\varphi_1 \vee \varphi_2$ in that case? According to the standard semantics this is determined entirely by the consistency constraints imposed by the axioms. For the function that has $P(\varphi_1) = 0.2$ and $P(\varphi_2) = 0.3$ we can derive $0.3 \leq P(\varphi_1 \vee \varphi_2) \leq 0.5$, and for the function that has $P(\varphi_2) = 0.4$ we can derive $0.4 \leq P(\varphi_1 \vee \varphi_2) \leq 0.6$. From these extreme cases we can conclude that $0.3 \leq P(\varphi_1 \vee \varphi_2) \leq 0.6$, so that $Y = [0.3, 0.6]$.

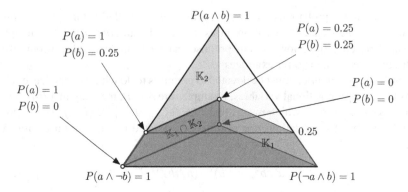

Fig. 2.2: The set of all possible probability measures P, depicted as a tetrahedron, together with the credal sets obtained from the given probabilistic constraints.

Example 2. Consider two premises $(a \wedge b)^{[0,0.25]}$ and $(a \vee \neg b)^1$. For the specification of a probability measure with respect to the corresponding 2-dimensional space $\{0, 1\}^2$, at least three parameters are needed (the size of the sample space minus 1). This means that the set of all possible probability measures P can be nicely depicted by a tetrahedron (3-simplex) with maximal probabilities for the state descriptions $a \wedge b$, $a \wedge \neg b$, $\neg a \wedge b$, and $\neg a \wedge \neg b$ at each of its four extremities. This tetrahedron is depicted in Figure 2.2, together with the convex sets $\mathbb{K}_1$ and $\mathbb{K}_2$ obtained from the constraints $P(a \wedge b) \in [0, 0.25]$ and $P(a \vee \neg b) = 1$, respectively. From the (convex) intersection $\mathbb{K}_1 \cap \mathbb{K}_2$, which includes all probability functions that satisfy both constraints, we see that $Y = [0, 1]$ attaches to the conclusion a, whereas $Y = [0, 0.25]$ attaches to the conclusion b.

Chapter 3
Probabilistic Argumentation

Degrees of support and possibility are the central formal concepts in the theory of *probabilistic argumentation* (Haenni, 2005a, 2009; Haenni et al., 2000; Kohlas, 2003). This theory is driven by the general idea of putting forward the pros and cons of a proposition or hypothesis in question. The weights of the resulting logical arguments and counter-arguments are measured by *probabilities*, which are then turned into (sub-additive[1]) degrees of *support* and (super-additive) degrees of *possibility*. Intuitively, degrees of support measure probabilistically the presence of evidence which supports the hypothesis, whereas degrees of possibility measure the absence of evidence which refutes the hypothesis. For this, probabilistic argumentation is concerned with probabilities of a particular type of event of the form 'the hypothesis is a logical consequence of the evidence' rather than 'the hypothesis is true', i.e., very much like Ruspini's *epistemic probabilities* (Ruspini, 1986; Ruspini et al., 1992). Apart from that, they are classical (additive[1]) probabilities in the sense of Kolmogorov's axioms.

Probabilistic argumentation as a computational process of formal reasoning has two major components. While the *qualitative* component deals with the generation of logical arguments and counter-arguments, it is up to the *quantitative* component to turn the qualitative results into numerical degrees of support and possibility. In the following overview, the focus will be placed on the quantitative component and the numerical results thereof. For a more comprehensive introduction and exposition of probabilistic argumentation we refer to (Haenni, 2009).

At this point, it should be mentioned that there is a strict mathematical analogy between degrees of support/possibility and belief/plausibility in the Dempster-Shafer theory of evidence (Dempster, 1968; Shafer, 1976). This connection has been thoroughly discussed in (Haenni and Lehmann, 2003), according to which any probabilistic argumentation system is expressible as a belief function. On the other hand,

---

[1] Degrees of support are additive probabilities with respect to the events that a given hypothesis is a logical consequence of the available evidence or not. However, by considering evidence and hypotheses which outrange the domain of the given probability space, they become sub-additive with respect to the hypothesis and its negation. This simple but remarkable effect is one of the theory's key components.

R. Haenni et al., *Probabilistic Logics and Probabilistic Networks*, Synthese Library 350,   21
DOI 10.1007/978-94-007-0008-6_3, © Springer Science+Business Media B.V. 2011

it is also possible to express belief functions as respective probabilistic argumentation systems and to formulate Dempster's combination rule as a particular form of
merging two probabilistic argumentation systems. Despite these technical similari-
ties, the theories are still quite different from a conceptual point of view. A major
difference is Dempster's rule of combination, which is a central conceptual element
in the Dempster-Shafer theory to combine different pieces of evidence, but which is
of almost no relevance in probabilistic argumentation. Note that the independence
assumption, on which Dempster's rule is based, has raised quite some criticism with
regard to the appropriateness of the rule and the theory as a whole (Zadeh, 1979).
In probabilistic argumentation, these criticisms are circumvented by not explicitly
formulating Dempster's rule and thus by not giving it such a fundamental role.

Another major difference is the fact that the notions of belief and plausibility
in the Dempster-Shafer theory are often entirely detached from a probabilistic interpretation (for example in Smets' *Transferable Belief Model* (Smets and Kennes,
1994)), whereas degrees of support and possibility are probabilities by definition.
Non-probabilistic interpretations have also raised many criticisms, e.g. by Pearl
(1990a), but they are no longer applicable within the pure probabilistic interpretation of probabilistic argumentation.

Finally, it should be noted that the use of logic to express factual information is
intrinsic in the theory of probabilistic argumentation, both as a powerful modeling
language and for computational purposes, but these ideas are almost inexistent in
the Dempster-Shafer theory.

3.1 Background

Probabilistic argumentation requires the available evidence to be encoded by a fi-
nite set $\Phi = \{\varphi_1, \ldots, \varphi_n\} \subset \mathscr{L}_V$ of sentences in a logical language $\mathscr{L}_V$ over a set of
variables V and a fully specified probability measure $P : 2^{\Omega_W} \to [0,1]$, where Ω_W
denotes the finite sample space generated by a subset $W \subseteq V$ of so-called *probabilistic variables*.[2] These are the theory's basic ingredients. The logical language
$\mathscr{L}_V$ itself is supposed to possess a well-defined model-theoretic semantics, in which
a monotonic (and decomposable) entailment relation $\models$ is defined in terms of set inclusion of models in some underlying universe Ω_V. Otherwise, there are no further
assumptions or restrictions regarding the logical language or the specification of the
probability measure P (for the latter we may for example use a Bayesian network).

Definition 4 (Probabilistic Argumentation System). A *probabilistic argumentation system* is a quintuple

$$\mathscr{A} = (V, \mathscr{L}_V, \Phi, W, P), \tag{3.1}$$

---

[2] The finiteness assumption with regards to Ω_W is not a conceptual restriction of the theory, but it
allows us here to define P with respect to the σ-algebra 2^{Ω_W} and thus helps to keep the mathematics
simple.

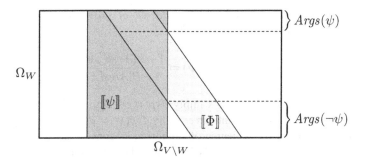

Fig. 3.1: The sample space Ω_W is shown as the vertical sub-space of the complete space $\Omega_V = \Omega_W \times \Omega_{V \setminus W}$, which means that the sets of arguments $Args(\psi)$ is the horizontal projection of the area of $[\![\Phi]\!]$ that lies entirely inside $[\![\psi]\!]$, whereas the set of counter-arguments $Args(\neg\psi)$ is the horizontal projection of the area of $[\![\Phi]\!]$ that lies entirely outside $[\![\psi]\!]$.

where V, $\mathcal{L}_V$, Φ, W, and P are as defined above (Haenni, 2009).

For a given probabilistic argumentation system $\mathcal{A}$, let another logical sentence $\psi \in \mathcal{L}_V$ represent the hypothesis in question. For the formal definition of degrees of support and possibility, consider the subset of Ω_W whose elements, if assumed to be true, are each sufficient to make ψ a logical consequence of Φ. Formally, this set of so-called *arguments* of ψ is defined by

$$Args_{\mathcal{A}}(\psi) = \{\omega \in \Omega_W : \Phi|\omega \models \psi\}, \tag{3.2}$$

where $\Phi|\omega$ denotes the set of sentences obtained from Φ by instantiating all the variables from W according to the partial truth assignment $\omega \in \Omega_W$ (Haenni, 2009). The elements of $Args_{\mathcal{A}}(\neg\psi)$ are sometimes called *counter-arguments* of ψ, see Figure 3.1 for an illustration. Note that the elements of $Args_{\mathcal{A}}(\bot) = Args_{\mathcal{A}}(\psi) \cap Args_{\mathcal{A}}(\neg\psi)$ are the ones that are inconsistent with the available evidence Φ, which is why they are called *conflicts*. The complement of the set of conflicts,

$$E_{\mathcal{A}} = \Omega_W \setminus Args_{\mathcal{A}}(\bot) = \{\omega \in \Omega_W : \Phi|\omega \not\models \bot\}, \tag{3.3}$$

can thus be interpreted as the available *evidence* in the sample space Ω_W induced by Φ. We will thus use $E_{\mathcal{A}}$ in its typical role to condition P. In the following, when no confusion is anticipated, we omit the reference to $\mathcal{A}$ and write E as a short form of $E_{\mathcal{A}}$ and $Args(\psi)$ as a short form of $Args_{\mathcal{A}}(\psi)$.

Definition 5 (Degree of Support). The *degree of support* of ψ, denoted by $dsp_{\mathcal{A}}(\psi)$ or simply by $dsp(\psi)$, is the conditional probability of the event $Args(\psi)$ given the evidence E,

$$dsp(\psi) = P(Args(\psi)|E) = \frac{P(Args(\psi)) - P(Args(\bot))}{1 - P(Args(\bot))}. \tag{3.4}$$

Definition 6 (Degree of Possibility). The *degree of possibility* of ψ, denoted $dps_{\mathscr{A}}(\psi)$ or simply by $dps(\psi)$, is defined by

$$dps(\psi) = 1 - dsp(\neg\psi) = \frac{1 - P(Args(\neg\psi))}{1 - P(Args(\bot))}. \tag{3.5}$$

Note that these formal definitions imply $dsp(\psi) \leq dps(\psi)$ for all hypotheses $\psi \in \mathscr{L}_V$, but this inequality turns into an equality $dsp(\psi) = dps(\psi)$ for all $\psi \in \mathscr{L}_W$. Another important property of degree of support is its consistency with pure logical and pure probabilistic inference. By looking at the extreme cases of $W = \emptyset$ and $W = V$, it turns out that degrees of support naturally degenerate into classical logical entailment $\Phi \models \psi$ and into ordinary posterior probabilities $P(\psi|\Phi)$, respectively. This underlines the theory's claim to be a unified formal theory of logical and probabilistic reasoning (Haenni, 2005a).

From a computational point of view, we can derive from Φ and ψ logical representations of the sets $Args(\psi)$, $Args(\neg\psi)$, and $Args(\bot)$ through *quantifier elimination* (Williams, 2002). For example, if the variables to be eliminated, $U = V \setminus W$, are all propositional variables, and if Φ is a clausal set, then it is possible to realize quantifier elimination as a resolution-based variable elimination procedure (Haenni et al., 2000).

Example 3. Consider a set $V = \{A_1, A_2, A_3, A_4, X, Y, Z\}$ of propositional variables with $W = \{A_1, A_2, A_3, A_4\}$ and therefore $U = \{X, Y, Z\}$. Furthermore, let $\Phi = \{a_1 \wedge a_2 \rightarrow x, a_3 \rightarrow y, x \vee y \rightarrow z, a_4 \rightarrow \neg z\}$ be the encoded evidence and $\psi = z$ the hypothesis in question. By eliminating the variables U from $\Phi \cup \{\neg z\}$ and $\Phi \cup \{z\}$ (using classical resolution-based variable elimination), we obtain

$$Args(z) = [\![(\neg a_1 \vee \neg a_2) \wedge \neg a_3]\!]^c = [\![(a_1 \wedge a_2) \vee a_3]\!],$$
$$Args(\neg z) = [\![\neg a_4]\!]^c = [\![a_4]\!],$$

respectively, which implies

$$Args(\bot) = Args(z) \cap Args(\neg z) = [\![((a_1 \wedge a_2) \vee a_3) \wedge a_4]\!].$$

If we suppose that the variables in W are probabilistically independent with respective marginal probabilities $P(a_1) = 0.7$, $P(a_2) = 0.2$, $P(a_3) = 0.5$, and $P(a_4) = 0.1$, we can use the induced probability measure P to obtain

$$P(Args(z)) = 0.57, \; P(Args(\neg z)) = 0.1, \; P(Args(\bot)) = 0.057,$$

from which we derive $dsp(z) = \frac{0.57 - 0.07}{1 - 0.057} = 0.544$ and $dps(z) = \frac{1 - 0.1}{1 - 0.057} = 0.954$. These results indicate the presence of some non-negligible arguments for z and the almost perfect absence of corresponding counter-arguments. In other words, there are some good reasons to accept, but almost no reason to reject z.

When it comes to quantitatively judging the truth of a hypothesis ψ, it is possible to interpret degrees of support and possibility as respective lower and upper bounds of a corresponding credal set. The fact that such bounds are obtained without effectively dealing with probability sets or probability intervals distinguishes the theory from most other approaches to probabilistic logic. Another important interpretation of degrees of support and possibility arises from seeing them as the coordinates $b = dsp(\psi)$, $d = 1 - dps(\psi)$, and $i = dps(\psi) - dsp(\psi)$ of an *opinion* $\omega_\psi = (b, d, i)$ in the standard 2-simplex $\{(b, d, i) \in [0, 1]^3 : b + d + i = 1\}$ called an *opinion triangle* (Jøsang, 1997; Haenni, 2009).

3.2 Representation

To connect probabilistic argumentation with probabilistic logic, let us first discuss a possible way of representing a probabilistic argumentation system $\mathscr{A} = (V, \mathscr{L}_V, \Phi, W, P)$ in form of the general framework of §1.2, where the available evidence is encoded in the form of Schema (1.1), i.e., as a set of sentences $\varphi_i^{X_i}$ with attached probabilistic weights $X_i \subseteq [0, 1]$.

The most obvious part of such an encoding are the sentences $\varphi_i \in \Phi$, which are all hard constraints with respect to the possible true state of Ω_V. To translate such model-theoretic constraints into corresponding probabilistic constraints, we simply attach the sharp value 1.0, or more strictly spoken the singleton set $X_i = \{1.0\}$, to each sentence φ_i. We will therefore have $\varphi_1^{1.0}, \ldots, \varphi_n^{1.0}$ as part of the left hand side of Schema (1.1), where 1.0 is a short form for the singleton set $\{1.0\}$.

The second part of the information contained in a probabilistic argumentation system is the probability measure $P : 2^{\Omega_W} \to [0, 1]$. The simplest and most general encoding in the form of Schema (1.1) consists in enumerating all elementary outcomes $\omega \in \Omega_W$ together with their respective probabilities $P(\{\omega\})$. For this, let $\alpha_\omega = [A_1 = e_1] \wedge \cdots \wedge [A_r = e_r]$ denote a conjunction, which assigns according to $\omega = (a_1^{e_1}, \ldots, a_r^{e_r})$ a value e_i to each of the r variables $A_1, \ldots, A_r \in W$. This leads to $[\![\alpha_\omega]\!] = \{\omega\}$ and thus allows us to use the sentence α_ω as a logical representative of ω. In the finite case, which means that the elements of $\Omega_W = \{\omega_1, \ldots, \omega_m\}$ are indexable, say from 1 to m, we finally obtain

$$\varphi_1^{1.0}, \ldots, \varphi_n^{1.0}, \alpha_{\omega_1}^{x_1}, \ldots, \alpha_{\omega_m}^{x_m}, \text{ with } x_i = P(\{\omega_i\}),$$

for a complete (but obviously not very compact) encoding of the probabilistic argumentation system in form of the left hand side of Schema (1.1). Note that all attached probabilities are sharp values, but depending of the chosen interpretation, this does not necessarily mean that the target set Y for a given conclusion ψ is also a sharp value (e.g., the standard semantics from §2 does not generally produce sharp values in such cases).

In case the probability measure P is specified in terms of marginal or conditional probabilities together with respective independence assumptions, for example by means of a Bayesian network, we would certainly not want to enumerate all atomic states individually. Instead we would rather try to express the given (conditional or marginal) probability values and independence assumptions *directly* by statements of the form of Schema (1.1). We have already seen in §2.2 how to represent conditional independence relations such as $A \perp\!\!\!\perp B|C$ in form of Schema (1.1), so we do not need to repeat the details of this technique at this point. In the particular case where all probabilistic variables are pairwise independent, we would have to add the constraints $A_i \perp\!\!\!\perp A_j$ for all pairs of probabilistic variables $A_i, A_j \in W$, $A_i \neq A_j$, together with respective constraints for their marginal probabilities $P(A_i)$.

3.3 Interpretation

Now let us move our attention to the question of interpreting instances of Schema (1.1) as respective probabilistic argumentation systems. For this, we will first generalize in four different ways the idea of the standard semantics as exposed in §2 to degrees of support and possibility. And then we will explore three different perspectives obtained by considering each attached probability set as an indicator of the premise's reliability. In all cases we will end up with lower and upper bounds for the target set Y on the right hand side of Schema (1.1). See (Haenni et al., 2008) for a related discussion.

3.3.1 Generalizing the Standard Semantics

As in the standard semantics, let the attached probability sets be interpreted as constraints on the possible probability measure P. We will see below that this can be done in various ways, but what these ways have in common is that the main components of the involved probabilistic argumentation system need to be fixed to get started. For this, let us first split up the set of premises into the ones with an attached probability of 1.0 and the ones with an attached probability or probability set different from 1.0. By taking the former for granted, the idea is to let them play the role of the available evidence Φ.

This decomposition of the set of premises is the common starting point of what follows, but to simplify the subsequent discussion and to make it most consistent with the rest of the book, let us simply assume Φ to be given *in addition* to some premises $\varphi_1^{X_1}, \ldots, \varphi_n^{X_n}$ in the form of Schema (1.1). If we then fix W to be the set of variables appearing in $\varphi_1, \ldots, \varphi_n$, we can apply the standard semantics to obtain the set

$$\mathbb{P} = \{P : P(\varphi_i) \in X_i, \forall i = 1, \ldots, n\}$$

of all admissible probability measures w.r.t. to the sample space Ω_W. The result is what could be called an *imprecise probabilistic argumentation system* $\mathscr{A} = (V, \mathscr{L}_V, \Phi, W, \mathbb{P})$. Or we may look at each probability measure from $\mathbb{P}$ individually and consider the family $\mathbb{A} = \{(V, \mathscr{L}_V, \Phi, W, P) : P \in \mathbb{P}\}$ of all such probabilistic argumentation systems, each of which with its own degree of support (and degree of possibility) function. Note that by applying this procedure to the proposed representation of the previous section, we return to the original probabilistic argumentation system (then both $\mathbb{P}$ and $\mathbb{A}$ degenerate into singletons).

Instead of using the sets X_i as constraints directly for P, we may also interpret them as respective constraints for corresponding degrees of support or possibility. This leads to the following three variations of the above scheme.

Constraints on Degrees of Support

If we consider each set X_i to be a constraint for the degree of support of φ_i, we obtain a set admissible probability measures that is quite different from the one above:

$$\mathbb{P} = \{P : dsp_{\mathscr{A}}(\varphi_i) \in X_i, \forall i = 1,\ldots,n, \ \mathscr{A} = (V, \mathscr{L}_V, \Phi, W, P)\}.$$

As before, this delivers a whole family $\mathbb{A} = \{(V, \mathscr{L}_V, \Phi, W, P) : P \in \mathbb{P}\}$ of possible probabilistic argumentation systems.

Constraints on Degrees of Possibility

In a similar way, we may consider each sets X_i to be a constraint for the degree of possibility of φ_i. The resulting set of admissible probability measures,

$$\mathbb{P} = \{P : dps_{\mathscr{A}}(\varphi_i) \in X_i, \forall i = 1,\ldots,n, \ \mathscr{A} = (V, \mathscr{L}_V, \Phi, W, P)\},$$

is again quite different from the ones above. Note that we may 'simulate' this semantics by applying the previous semantics to the negated premises $\neg\varphi_1^{Z_1}, \ldots, \neg\varphi_n^{Z_n}$, where $Z_i = \{1 - x : x \in X_i\}$ denotes the corresponding 'negated' sets of probabilities, and the same works in the other direction. This remarkable relationship is a simple consequence of the duality between degrees of support and possibility.

Combined Constraints

To obtain a more symmetrical semantics, in which degrees of support and degrees of possibility are equally important, we consider the restricted case where each set $X_i = [\ell_i, u_i]$ is an interval. We may then interpret the lower bound ℓ_i as a sharp constraint for the degree of support and the upper bound u_i as a sharp constraint for the degree of possibility of φ_i. This defines another set of admissible probability measures,

$$\mathbb{P} = \{P : dsp_{\mathscr{A}}(\varphi_i) = \ell_i, dps_{\mathscr{A}}(\varphi_i) = u_i, \forall i = 1,\ldots,n, \ \mathscr{A} = (V, \mathscr{L}_V, \Phi, W, P)\},$$

which is again quite different from the previous ones. Note that we can use the relationship $dps(\psi) = 1 - dsp(\neg\psi)$ to turn the constraints $dsp(\psi_i) = \ell_i$ and $dps(\psi_i) = u_i$ into two constraints for respective degrees of support or into two constraints for respective degrees of possibility, whatever is more desirable.

To use any of those interpretations to produce an answer to our main question regarding the extent of the set Y for a conclusion ψ, there are again different ways to proceed, depending on whether degrees of support or degrees of possibility are of principal interest.

1. The first option is to consider the target set $Y_{dsp} = \{dsp_{\mathscr{A}}(\psi) : \mathscr{A} \in \mathbb{A}\}$, which consists of all possible degrees of support w.r.t $\mathbb{A}$.
2. As a second option, we may do the same with degrees of possibility, from which we get another possible target set $Y_{dps} = \{dps_{\mathscr{A}}(\psi) : \mathscr{A} \in \mathbb{A}\}$.
3. Finally, we may want to look at the minimal degree of support, $\underline{dsp}(\psi) = \min\{dsp_{\mathscr{A}}(\psi) : \mathscr{A} \in \mathbb{A}\}$, as well as the maximal degree of possibility, $\overline{dps}(\psi) = \max\{dps_{\mathscr{A}}(\psi) : \mathscr{A} \in \mathbb{A}\}$, and use them as respective lower and upper bounds for the target interval $Y_{dsp/dps} = [\underline{dsp}(\psi), \overline{dps}(\psi)]$. By doing so, we depart from the general idea of the standard semantics that the target interval is a set of probabilities satisfying the given constraints. However, we may still consider it a useful interpretation for instances of Schema (1.1).

Notice that in the special case of $\Phi = \emptyset$, which implies $W = V$, all three options coincide with the standard semantics from §2.

3.3.2 Premises from Unreliable Sources

Some very simple, but quite different semantics arise when X_i is used to express the *evidential uncertainty* of the premise φ_i in the sense that φ_i belongs to Φ with probability $x_i \in X_i$. Such situations may appear from collecting the premises from various unreliable sources. Formally, we could express this idea by $P(\varphi_i \in \Phi) \in X_i$ and thus interpret Φ as an 'imprecise fuzzy set' whose membership function is only partially determined by the attached probability set.

This way of looking at questions in the form of Schema (1.1) again allows various concrete interpretations. Three options will be discussed in the remaining of this section. In each case, we will end up with sharp degrees of support and possibility, which will be used as respective lower and upper bounds for the target interval Y. Note that this is again quite different from the general idea of the standard semantics, where the target interval is a set of probabilities satisfying the given constraints, whereas here the interval itself is not a quantity, only the bounds are quantities. Apart from that, we may still consider them as useful interpretations for instances of Schema (1.1).

3.3.2.1 Incompetent Sources

Suppose that each of the available premises has a sharp probability $X_i = \{x_i\}$ attached to it. To make this setting compatible with a probabilistic argumentation system, let us first redirect each attached probability x_i to an auxiliary propositional variable REL_i. The intuitive idea of this is to consider each premise φ_i as a piece of evidence from a possibly unreliable source S_i. The *reliability* of S_i is thus modeled by the proposition rel_i (which abbreviates $REL_i = true$), and with $P(rel_i) = x_i$ we measure its degree of reliability. The subsequent discussion will be restricted to the case of *independent*[3] sources, which allows us to multiply the marginal probabilities $P(REL_i)$ to obtain a fully specified probability measure P over all auxiliary variables.

On the purely logical side, we should expect that any statement from a reliable source is indeed true. This allows us to write $rel_i \rightarrow \varphi_i$ to connect the auxiliary variable REL_i with φ_i. With

$$\Phi^+ = \{rel_1 \rightarrow \varphi_1, \dots, rel_n \rightarrow \varphi_n\}$$

we denote the set of all such material implications, from which we obtain a probabilistic argumentation system $\mathscr{A}^+ = (V \cup W, \mathscr{L}_{V \cup W}, \Phi^+, W, P)$ with $W = \{REL_1, \dots, REL_n\}$ and P as defined above. This allows us then to compute the degrees of support and possibility for the conclusion ψ and to use them as lower and upper bounds for the target interval Y.

In the proposed setting, only the positive case of a reliable source is modeled, but nothing is said about the behaviour of an unreliable source. For this, it is possible to distinguish between *incompetent* and *dishonest* (but competent) sources. In the case of an incompetent source, from which no meaningful evidence should be expected, we may model the negative behaviour by auxiliary implications of the form $\neg rel_i \rightarrow \top$. Note that these implications are all irrelevant tautologies, i.e., we get back to the same set Φ^+ from above. In this semantics, the values $P(rel_i) = x_i$ should therefore be interpreted as *degrees of competence* rather than general degrees of reliability.

Example 4. Let $V = \{X, Y, Z\}$ be a set of propositional variables and

$$x^{0.7}, x \rightarrow y^{0.9}, \neg y \leftrightarrow z^{0.3} \approx z^Y$$

the given problem in the form of Schema (1.1). Using the necessary auxiliary variables $W = \{REL_1, REL_2, REL_3\}$ with respective marginal probabilities $P(rel_1) = 0.7$, $P(rel_2) = 0.9$, and $P(rel_3) = 0.3$, we obtain the set

$$\Phi^+ = \{rel_1 \rightarrow x, \ rel_2 \rightarrow (x \rightarrow y), \ rel_3 \rightarrow (\neg y \leftrightarrow z)\},$$

---

[3] This assumption may appear to be overly idealized, but there are many practical situations in which this is approximately correct (Haenni, 2005b; Haenni and Hartmann, 2006). Relaxing the independence assumption would certainly allow us to cover a broader class of problems, but it would also make the analysis more complicated.

and a corresponding probabilistic argumentation system $\mathscr{A}^+$. For the hypothesis z, this leads to the $dsp_{\mathscr{A}^+}(z) = 0$ and $dps_{\mathscr{A}^+}(z) = 0.811$ and thus to $Y = [0, 0.811]$ for the resulting target interval Y of Schema (1.1).

Dishonest Sources

As before, we suppose that all attached probabilities are sharp values x_i, but now we consider the possibility of the sources being *malicious*, i.e., competent but not necessarily honest. In this case, the interpretation of $P(rel_i) = x_i$ becomes the one of a *degree of honesty* of source S_i. Dishonest sources are different from incompetent sources in their attitude of deliberately stating the opposite of the truth. From a logical point of view, $\neg rel_i$ allows us thus to infer $\neg \varphi_i$, which we may express by additional material implications $\neg rel_i \rightarrow \neg \varphi_i$. This leads to an extended set of premises,

$$\Phi^{\pm} = \Phi^+ \cup \{\neg rel_1 \rightarrow \neg \varphi_1, \ldots, \neg rel_n \rightarrow \neg \varphi_n\} \equiv \{rel_1 \leftrightarrow \varphi_1, \ldots, rel_n \leftrightarrow \varphi_n\},$$

and a new probabilistic argumentation system $\mathscr{A}^{\pm} = (V \cup W, \mathscr{L}_{V \cup W}, \Phi^{\pm}, W, P)$. Note that the difference between the two interpretations may have a huge impact on the resulting degrees of support and possibility of ψ, and therefore produce quite different target sets Y, as demonstrated in the following example.

Example 5. Consider the setting from Example 4, from which we get an extended set of premises,

$$\Phi^{\pm} = \{rel_1 \leftrightarrow x, \, rel_2 \leftrightarrow (x \rightarrow y), \, rel_3 \leftrightarrow (\neg y \leftrightarrow z)\},$$

and a new probabilistic argumentation system $\mathscr{A}^{\pm} = (V \cup W, \mathscr{L}_{V \cup W}, \Phi^{\pm}, W, P)$. This leads then to $dsp_{\mathscr{A}^{\pm}}(z) = 0.476$ and $dps_{\mathscr{A}^{\pm}}(z) = 0.755$, and finally we obtain $Y = [0.476, 0.755]$ for the resulting target interval.

Incompetent and Dishonest Sources

In the more general case, where each $X_i = [\ell_i, u_i]$ is an interval, we will now consider a refined model of the above-mentioned idea of splitting up reliability into competence and honesty. Let X_i still refer to the reliability of the source, but consider now two auxiliary variables $COMP_i$ (for competence) and HON_i (for honesty). This allows us to distinguish three exclusive and exhaustive cases, namely $comp_i \wedge hon_i$ (the source is reliable), $comp_i \wedge \neg hon_i$ (the source is malicious), and $\neg comp_i$ (the source is incompetent). As before, we assume that φ_i holds if S_i is reliable, but also that $\neg \varphi_i$ holds if S_i is malicious. Statements from incompetent sources will again be neglected. Logically, the general behaviour of such a source can thus be modeled by two sentences $comp_i \wedge hon_i \rightarrow \varphi$ and $comp_i \wedge \neg hon_i \rightarrow \neg \varphi_i$, which can be merged into $comp_i \rightarrow (hon_i \leftrightarrow \varphi_i)$. This leads to the set of premises

$$\Phi^* = \{comp_1 \rightarrow (hon_1 \leftrightarrow \varphi_1), \ldots, comp_n \rightarrow (hon_n \leftrightarrow \varphi_n)\}.$$

To turn this model into a probabilistic argumentation system, we need to link the auxiliary variables $W = \{COMP_1, \ldots, COMP_n, HON_1, \ldots, HON_n\}$ to corresponding probabilities. For this, we assume independence between $COMP_i$ and HON_i, which is often quite reasonable. If we assume the least restrictive interval $X_i = [0,1]$ to represent a totally incompetent source, and similarly the most restrictive interval $X_i = [x_i, x_i]$ to represent a totally competent source, then $u_i - \ell_i$ surely represents the source's degree of incompetence, from which we obtain

$$P(comp_i) = 1 - (u_i - \ell_i) = 1 - u_i + \ell_i$$

for the marginal probability of $comp_i$. Following a similar line of reasoning, we first obtain $P(comp_i \wedge hon_i) = \ell_i$ for the combined event $comp_i \wedge hon_i$ of a reliable source, which then leads to

$$P(hon_i) = \frac{\ell_i}{P(comp_i)} = \frac{\ell_i}{1 - u_i + l_i}$$

for the marginal probability of hon_i. As before, we can use the independence assumption to multiply these values to obtain a fully specified probability measure P over all auxiliary variables. With $\mathscr{A}^* = (V \cup W, \mathscr{L}_{V \cup W}, \Phi^*, W, P)$ we denote the resulting probabilistic argumentation system, from which we obtain degrees of support and possibility for ψ, the bounds for the target interval Y. Note that $\mathscr{A}^+$ and $\mathscr{A}^{\pm}$ from the previous two semantics are special cases of $\mathscr{A}^*$, namely for $u_i = 1$ (hon_i becomes irrelevant, and rel_i undertakes the role of $comp_i$) and $\ell_i = u_i$ ($comp_i$ becomes irrelevant, and rel_i undertakes the role of hon_i), respectively.

Example 6. Consider the following adapted version of the problem in Example 5, in which the original sharp values are replaced by intervals:

$$x^{[0.6,0.8]}, x \rightarrow y^{[0.9,1]}, \neg y \leftrightarrow z^{[0.1,0.5]} \approx z^Y.$$

For $W = \{COMP_1, COMP_2, COMP_3, HON_1, HON_2, HON_3\}$, this implies the following marginal probabilities:

$$P(comp_1) = 0.8, \qquad P(comp_2) = 0.9, \qquad P(comp_3) = 0.6$$
$$P(hon_1) = 0.75, \qquad P(hon_2) = 1, \qquad P(hon_3) = 0.2.$$

Together with the proposed encoding and the resulting set of sentences,

$$\Phi^* = \{comp_1 \rightarrow (hon_1 \leftrightarrow x), comp_2 \rightarrow (hon_2 \leftrightarrow (x \rightarrow y)),$$
$$comp_3 \rightarrow (hon_3 \leftrightarrow (\neg y \leftrightarrow z))\},$$

we finally obtain $dsp_{\mathscr{A}^*}(z) = 0.216$ and $dps_{\mathscr{A}^*}(z) = 0.946$, which implies the target interval $Y = [0.216, 0.946]$.

Chapter 4
Evidential Probability

Rudolf Carnap (1962) distinguished between probability$_1$, which concerns rational degrees of belief, and probability$_2$, which concerns statistical regularities. Although he claimed that both notions of probability were crucial to scientific inference, he practically ignored probability$_2$ in the development of his systems of inductive logic. By contrast, evidential probability, developed by Henry Kyburg (1961) and later by Kyburg and Choh Man Teng (2001), is a theory that gives primacy to probability$_2$, and Kyburg's philosophical program was an uncompromising approach to see how far he could go with relative frequencies.

EP is motivated by two simple ideas, which together have not so simple consequences: probability assessments should be based upon relative frequencies, to the extent that we know them, and the assignment of probability to specific events should be determined by everything that is known about that specific event. Regarding the first point, evidential probability assignments are made *only* on the basis of observed relative frequencies. So, since we rarely observe a joint distribution, it should not be not surprising that EP inference is not a form of Bayesian inference. Regarding the second point, you may have knowledge about a specific event that links it to conflicting statistical information. A consequence of EP methods for resolving conflict is that EP inference may conflict with Bayesian methods (Kyburg, 1974; Levi, 1977; Kyburg, 2007; Seidenfeld, 2007). Even so, we'll see that there is a place for a version of evidential probability within our framework.

4.1 Background

Evidential probability is fundamentally a logic of probability rather than a probabilistic logic (Levi, 2007), where the evidential probability of χ given Γ_δ, $\mathrm{Prob}(\chi, \Gamma_\delta)$, is not a ratio but a 2-place meta-linguistic operator on analogy with provability. A good introduction to EP is (Kyburg and Teng, 2001), and the collected papers in (Harper and Wheeler, 2007) include both recent developments as well as critical essays.

R. Haenni et al., *Probabilistic Logics and Probabilistic Networks*, Synthese Library 350, 33
DOI 10.1007/978-94-007-0008-6_4, © Springer Science+Business Media B.V. 2011

The semantics governing the operator $\text{Prob}(\cdot,\cdot)$ is dissimilar to axiomatic theories of probability that take conditional probability as primitive, such as the system developed by Lester Dubbins (1975; Arló-Costa and Parikh, 2005), and it also resists reduction to linear (de Finetti, 1974) as well as to lower previsions (Walley, 1991). One difference between EP and the first two theories is that EP is interval-valued rather than point-valued, because the relative frequencies that underpin assignment of evidential probability are typically incomplete and approximate. But more generally, EP assignments may violate the coherence principles of imprecise probability theory. This point will become clearer when we discuss the principles for sorting relevant from irrelevant statistics.

Evidential probability is a logic of statistical probability statements and there is nothing in the activity of observing and recording statistical regularities that guarantees that a set of statistical probability statements will comport to the axioms of probability. So, EP is neither a species of Carnapian logical probability nor a kind of Bayesian probabilistic logic.[1,2] EP is instead a logic for approximate reasoning, thus it is more similar in kind to the theory of rough sets (Pawlak, 1991) and to systems of fuzzy logic (Dubois and Prade, 1980) than to probabilistic logic.

The operator $\text{Prob}(\cdot,\cdot)$ takes as arguments a sentence χ in the first coordinate and a set of statements Γ_δ in the second. The statements in Γ_δ represent a knowledge base, which includes categorical statements as well as statistical generalities. Theorems of logic and mathematics are examples of categorical statements, but so too are contingent generalities, such as the ideal gas law. EP treats the propositions '$2+2=4$' and '$PV = nRT$' (within a chemistry knowledge base, say) as indistinguishable analytic truths built into a particular language adopted for handling statistical statements to do with gasses.

Statistical generalities within Γ_δ, by contrast, are viewed as direct inference statements and are represented by syntax that is unique to evidential probability. *Direct inference* is the probability assigned to a target subclass given known frequency information about a reference population, and is often contrasted to *indirect inference*, which is the assignment of probability to a population given observed frequencies in a sample. Kyburg's ingenious idea was to solve the problem of indirect inference by viewing it as a form of direct inference.

Direct inference statements are statements that record the observed frequency of items satisfying a specified reference class that also satisfy a particular target class, and take the form of

$$\%\mathbf{x}(\tau(\mathbf{x}), \rho(\mathbf{x}), [l, u]).$$

---

[1] See the essays by Levi, Seidenfeld in (Harper and Wheeler, 2007) for a discussion of the sharp differences between EP and Bayesian approaches, particularly on the issue of conditionalization. A point sometimes overlooked by critics is that there are different *systems* of evidential probability corresponding to different conditions we assume to hold. Results pertaining to a qualitative representation of EP inference, for instance, assumes that Γ_δ is consistent. A version of conditionalization holds in EP given that there is specific statistical statement pertaining to the relevant joint distribution. See (Kyburg, 2007) and (Teng, 2007).

[2] EP does inherit some notions from Keynes's, however, including that probabilities are interval-valued and not-necessarily comparable.

This schematic statement says that given a sequence of propositional variables $\mathbf{x}$ that satisfies the reference class predicate ρ, the proportion of ρ that also satisfies the target class predicate τ is between l and u.

Syntactically, '$\tau(\mathbf{x}), \rho(\mathbf{x}), [l, u]$' is an open formula schema, where '$\tau(\cdot)$' and '$\rho(\cdot)$' are replaced by open first-order formulas, '$\mathbf{x}$' is replaced by a sequence of propositional variables, and '$[l, u]$' is replaced by a specific sub-interval of $[0, 1]$. The binding operator '$\%$' is similar to the ordinary binding operators ($\forall, \exists$) of first-order logic, except that '$\%$' is a 3-place binding operator over the propositional variables appearing within the *target formula* $\tau(\mathbf{x})$ and the *reference formula* $\rho(\mathbf{x})$, and binds those formulas to an interval.[3] The language $\mathscr{L}^{ep}$ of evidential probability then is a guarded first-order language (Andréka et al., 1998) augmented to include direct inference statements. There are additional formation rules for direct inference statements that are designed to block spurious inference, but we shall pass over these details of the theory.[4] An example of a direct inference statement that might appear in Γ_δ is

$$\%x(B(x), A(x), [.71, .83]),$$

which expresses that the proportion of A's that are also B's lies between 0.71 and 0.83.

As for semantics, a model M of $\mathscr{L}^{ep}$ is a pair, $\langle \mathscr{D}, \mathscr{I} \rangle$, where $\mathscr{D}$ is a two-sorted domain consisting of mathematical objects, $\mathscr{D}_m$, and a *finite* set of empirical objects, $\mathscr{D}_e$. EP thus assumes that there is a first mongoose and a last carbon molecule. $\mathscr{I}$ is an interpretation function that is the union of two partial functions, one defined on $\mathscr{D}_m$ and the other on $\mathscr{D}_e$. Otherwise M behaves like a first-order model: the interpretation function $\mathscr{I}$ maps (empirical/mathematical) terms into the (empirical/mathematical) elements of $\mathscr{D}$, monadic predicates into subsets of $\mathscr{D}$, n-arity relation symbols into $\mathscr{D}^n$, and so forth. Variable assignments also behave as one would expect, with the only difference being the procedure for assigning truth to direct inference statements.

The basic idea behind the semantics for direct inference statements is that the statistical quantifier '$\%$' ranges over the finite empirical domain $\mathscr{D}_e$, not the field terms l, u that denote real numbers in $\mathscr{D}_m$. This means that the only free variables in a direct inference statement range over a finite domain, which will allow us to look at proportions of models in which a sentence is true. A *satisfaction set* of an open formula φ whose only free n variables are empirical is the subset of $\mathscr{D}^n$ that satisfies φ.

A direct inference statement $\%x(\tau(x), \rho(x), [l, u])$ is true in M under variable assignment v iff the cardinality of the satisfaction sets for the open formula ρ under v is greater than 0 and the ratio of the cardinality of satisfaction sets for $\tau(x^*) \wedge \rho(x^*)$ over the cardinality of the satisfaction sets for $\rho(x)$ (under v) is in the closed interval [l,u], where all variables of x occur in ρ, all variables of τ occur in ρ, and x^* is the sequence of variables free in ρ but not bound by $\%x$ (Kyburg and Teng, 2001).

---

[3] Hereafter we relax notation and simply use an arbitrary variable 'x' for '$\mathbf{x}$'.

[4] For details, see (Kyburg and Teng, 2001).

Now that we have a picture of what EP is, we turn to consider the inferential behavior of the theory. We propose to do this with a simple ball-draw experiment before considering the specifics of the theory in more detail in the next section. We then present an extended example in §4.1.2, which features more realistic data.

Example 7. Suppose the proportion of white balls (W) in an urn (U) is known to be within $[.33,4]$, and that ball t is drawn from U. These facts are represented in Γ_δ by the sentences, $\%x(W(x),U(x),[.33,.4])$ and $U(t)$.

1. If these two statements are all that we know about t, i.e., they are the only statements in Γ_δ pertaining to t, then $\mathrm{Prob}(W(t),\Gamma_\delta) = [.33,.4]$.
2. Suppose additionally that the proportion of plastic balls (P) that are white is observed to be between $[.31,.36]$, t is plastic, and that every plastic ball is an urn ball. That means that $\%x(P(x),U(x),[.31,.36])$, $P(t)$, and $\forall x.P(x) \rightarrow U(x)$ are added to Γ as well. Then there is conflicting statistical knowledge about o, since either:

   a. the probability that ball t is white is between $[.33,.4]$,
      by reason of $\%x(W(x),U(x),[.33,.4])$, or

   b. the probability that ball t is white is between $[.31,.36]$,
      by reason of $\%x(W(x),P(x),[.31,.36])$,

   may apply. There are several ways that statistical statements may conflict and there are rules for handling each type, which we will discuss in the next section. But in this particular case, because it is known that the class of plastic balls is more *specific* than the class of balls in U and we have statistics for the proportion of plastic balls that are also white balls, the statistical statement in (2) dominates the statement in (1). So, the probability that t is white is between $[.31,.36]$.
3. Adapting an example from (Kyburg and Teng, 2001, 216), suppose U is partitioned into three cells, u_1, u_2, and u_3, and that the following compound experiment is performed. First, a cell of U is selected at random. Then a ball is drawn at random from that cell. To simplify matters, suppose that there are 25 balls in U and 9 are white such that 3 of 5 balls from u_1 are white, but only 3 of 10 balls in u_2 and 3 of 10 in u_3 are white. The following table summarizes this information.

Table 4.1: Compound Experiment

	u_1	u_2	u_3	
W	3	3	3	9
$\overline{W}$	2	7	7	16
	5	10	10	25

We are interested in the probability that t is white, but we have a conflict. Given these over all precise values, we would have $\mathrm{Prob}(W(t),\Gamma_\delta) = \frac{9}{25}$. However, since we know that t was selected by performing this compound experiment, then we also have the conflicting direct inference statement $\%x,y(W^*(x,y),U^*(x,y), [.4,.4])$, where U^* is the set of compound two stage experiments, and W^* is the set of outcomes in which the ball selected is white.[5] We should prefer the statistics from the compound experiment because they are *richer* in information. So, the probability that t is white is .4.

4. Finally, if there happens to be *no* statistical knowledge in Γ_δ pertaining to t, then we would be completely ignorant of the probability that t is white. So in the case of total ignorance, $\text{Prob}(W(t),\Gamma_\delta) = [0,1]$.

4.1.1 Calculating Evidential Probability

In practice an individual may belong to several reference classes with known statistics. Selecting the appropriate statistical distribution among the class of potential probability statements is the *problem of the reference class* (Reichenbach, 1949). The task of assigning evidential probability to a statement χ relative to a set of evidential certainties relies upon a procedure for eliminating excess candidates from the set of potential candidates. This procedure is described in terms of the following definitions.

Definition 7 (Potential Probability Statement). A *potential probability statement* for χ with respect to Γ_δ is a tuple $\langle t, \tau(t), \rho(t), [l,u] \rangle$, such that instances of $\chi \leftrightarrow \tau(t)$, $\rho(t)$, and $\%x(\tau(x),\rho(x),[l,u])$ are each in Γ_δ.

Given χ, there are possibly many target statements of form $\tau(t)$ in Γ_δ that have the same truth value as χ. If it is known that individual t satisfies ρ, and known that between .7 and .8 of ρ's are also τ's, then $\langle t, \tau(t), \rho(t), [.7,.8] \rangle$ represents a potential probability statement for χ based on the knowledge base Γ_δ. Our focus will be on the statistical statements $\%x(\tau(x),\rho(x),[l,u])$ in Γ_δ that are the basis for each potential probability statement.

Selecting the appropriate probability interval for χ from the set of potential probability statements reduces to identifying and resolving conflicts among the statistical statements that are the basis for each potential probability statement. Thus the notion of conflict is central to the enterprise.

Two potential probability statements conflict if the probability interval of one is not a sub-interval of another.

Definition 8 (Conflict). Two intervals $[l,u]$ and $[l',u']$ *conflict* iff neither $[l,u] \subset [l',u']$ nor $[l,u] \supset [l',u']$. Two statistical statements conflict iff their intervals conflict.

So, conflicting intervals may be disjoint or intersect.

---

[5] Γ_δ should also include the categorical statements $\forall x,y(U^*\langle x,y\rangle \rightarrow W(y))$, which says that the second stage of U concerns the proportion of balls that are white, and three statements of the form $W^*(\mu,t) \leftrightarrow W(t)$, where μ is replaced by u_1,u_2,u_3, respectively. This statement tells us that everything that's true of W^* is true of W, which is what ensures that this conflict is detected.

Definition 9 (Cover). An interval $[l, u]$ *covers* a set of intervals $\mathscr{I}$ iff for every $[l', u'] \in \mathscr{I}$, $l \leq l'$ and $u' \leq u$. A cover $[l, u]$ of $\mathscr{I}$ is the *smallest cover* iff, for all covers $[l^*, u^*]$ of $\mathscr{I}$, $l^* \leq l$ and $u \leq u^*$.

A set of intervals $\mathscr{I}$ is closed under conflict if $\mathscr{I}$ contains no cover, and a set of potential statistical statements is closed under conflict if their intervals are closed under conflict.

Definition 10 (Closed Under Conflict). Let $\mathscr{I}$ be a non-empty set of intervals arranged in a sequence $\langle X_1, X_2, \ldots, X_n \rangle$, where l_i and u_i are the lower bound and the upper bound, respectively, of the ith interval X_i in $\mathscr{I}$. Define

(i) $L = \langle X_1', X_2', \ldots, X_n' \rangle$ to be a permutation of $\mathscr{I}$ such that for all $j > i$, either $l_i' < l_j'$ or $\left(l_i' = l_j' \text{ and } u_i' \geq u_j' \right)$.

(ii) $U = \langle X_1'', X_2'', \ldots, X_n'' \rangle$ to be a permutation of $\mathscr{I}$ such that for all $j > i$, either $u_i'' > u_j''$ or $\left(u_i'' = u_j'' \text{ and } l_i'' \leq l_j'' \right)$.

Then:

1. If $|\mathscr{I}| = 1$, the closure under conflict of $\mathscr{I} = \mathscr{I}$, i.e., $CC(\mathscr{I}) = \{X_1\}$.
2. Otherwise, define $S_0 = \{X_n', X_n''\}$ and the smallest cover of S_0 as $[l_0, u_0]$. Let $i = 0$.

   a. Then define S_{i+1} as the set $S_i \cup \{X_k \in \mathscr{I} :$ either the lower bound of X_k is greater than the lower bound of S_i, $l_k > l_i$, or the upper bound of X_k is less than the upper bound of S_i, $u_k < u_i\}$.
   b. Repeat until $S_{i+1} = S_i$, in which case denote by S.
   c. Then, $CC(\mathscr{I}) = S$.

Example 8. Let $\mathscr{I} = \{[.25, .35], [.4, .4], [.2, .45], [.3, .3], [.2, .5]\}$. Then, with respect to $\mathscr{I}$,

- $L = \langle [.2, .5], [.2, .45], [.25, .35], [.3, .3], [.4, .4] \rangle$, and
- $U = \langle [.2, .5], [.2, .45], [.4, .4], [.25, .35], [.3, .3] \rangle$.

The $CC(\mathscr{I}) = \{[.25, .35], [.4, .4], [.3, .3]\}$.

EP resolves conflicting statistical data concerning χ by applying two principles to the set of potential probability assignments, *Richness* and *Specificity*, to yield a class of *relevant statements*. The (controversial) principle of *Strength* is then applied to this set of relevant statistical statements, yielding a unique probability interval for χ. For discussion of these principles, see (Teng, 2007).

1. **[Richness]** If φ and ϑ conflict, ϑ is based on a marginal distribution while φ is based on a joint distribution, and φ cannot be eliminated by other applications of Richness, then eliminate $\langle \tau, \rho_\vartheta, [l, u] \rangle$ from the set of potential probability statements.
2. **[Specificity]** If φ and ϑ both survive the principle of richness, and if $\rho_\varphi \subset \rho_\vartheta$, then eliminate $\langle \tau, \rho_\vartheta, [l, u] \rangle$ from the set of potential probability statements.
3. **[Strength]** Let Γ^{RS} be the set of relevant statistical statements for χ with respect to Γ_δ, and let $\mathscr{I}$ be the corresponding set of relevant intervals based on Γ^{RS}. The principle of strength is the smallest cover of $CC(\mathscr{I})$.

The Principle of Richness says that a conflict between statistics based on a joint distribution and statistics based on a marginal should be resolved in favor of the joint distribution.

The Principle of Specificity says that if it is known that the reference class ρ_ϑ is included in the reference class ρ_φ, then eliminate the statement φ.

The statistical statements that survive the sequential application of Richness followed by Specificity are called *relevant statistics*. The Principle of Strength says to take the closure under conflict of the relevant statistics and identify the shortest cover of that set of intervals. That interval is the evidential probability assigned to the event.

Example 9. Suppose that a statement χ about individual t has these five potential probability statements:

1. $\langle t, A(t), P(t), [.25, .35] \rangle$
2. $\langle t, B(t), Q(t), [.4, .4] \rangle$
3. $\langle t, C(t), R(t), [.2, .25] \rangle$
4. $\langle t, D(t), S(t), [.3, .3] \rangle$
5. $\langle t, E(t), T(t), [.2, .5] \rangle$

Call this set of potential probability statements Γ. Now consider three different evidential situations involving Γ

- *Total knowledge*: If Γ is all that is known about χ and there is no additional information about the relationships between the statistical classes underlying these statements, then the set of conflicting statements is $\{1, 2, 4\}$, which is identical to $CC(\mathscr{I})$ in Example 8. Neither Richness nor Specificity succeeds in pruning away potential probability statements, since there is no additional information. So, Strength assigns the shortest cover, $[.25, .4]$, as the probability that χ
- *Specific information on reference classes*: Consider now the following two relationship among reference classes:

  a) $S \subsetneq Q$
  b) $Q \subsetneq S$

Consider the evidential situation in which a) is known. Then, the initial set of conflicting statements is $\{1, 2, 4\}$ as before. However, the clause a) allows for the Specificity rule to be applied, yielding the set $\{1, 4\}$. But notice that the corresponding set of intervals for this set is not closed under conflict:

$$CC(\{[.25, .35], [.3, .3]\}) \neq \{[.25, .35], [.3, .3]\}$$

Instead, $CC(\{[.25, .35], [.3, .3]\}) = \{[.3, .3]\}$. So, by strength (trivially), the probability is .3

Compare now the evidential situation in which b) is known instead. The initial set of conflicting statements is $\{1, 2, 4\}$, but the application of specificity removes $[.3, .3]$ in this case rather than $[.4, .4]$. But notice that in this case there is no change, that is

$$CC(\{[.25,.35],[.4,.4]\}) = CC(\mathscr{I})$$

So, by strength, the probability is $[.25,.4]$.

We may define Γ_ε, the set of practical certainties, in terms of a body of evidence Γ_δ:

$$\Gamma_\varepsilon = \{\chi : \exists\, l,u\,(\mathrm{Prob}(\neg\chi,\Gamma_\delta) = [l,u] \wedge u \leq \varepsilon)\},$$

or alternatively,

$$\Gamma_\varepsilon = \{\chi : \exists\, l,u\,(\mathrm{Prob}(\chi,\Gamma_\delta) = [l,u] \wedge l \geq 1 - \varepsilon)\}.$$

The set Γ_ε is the set of statements that the evidence Γ_δ warrants accepting; we say a sentence χ is *ε-accepted* if $\chi \in \Gamma_\varepsilon$. Thus we may add to our knowledge base statements that are non-monotonic consequences of Γ_δ with respect to a threshold point of acceptance.

The set Γ^{RS} contains all sets of undominated direct inference statements with respect to a particular sentence, χ. Corresponding to the relevant statistics for the pair $\langle\chi,\Gamma^{RS}\rangle$ are sets of satisfaction sets, which are called *support models* of Γ_δ for χ. EP is sound in the following sense: if $Prob(\chi,\Gamma_\delta) = [l,u]$, then the proportion of support models of Γ_δ in which χ is true lies between l and u (Kyburg and Teng, 2001). This result is the basis for EP's semantics for *partial entailment*.

Finally, we may view the evidence Γ_δ as providing real-valued bounds on 'degrees of belief' owing to the logical structure of sentences accepted into Γ_δ. However, the probability interval $[l,u]$ associated with χ does not specify a range of degrees of belief between l and u: the interval $[l,u]$ itself is not a quantity, only l and u are quantities, which are used to specify bounds. On this view, no degree of belief within $[l,u]$ is defensible, which is in marked contrast to the view offered in §7, Objective Bayesianism, which utilizes an entropic principle to assign degrees of belief (Wheeler and Williamson, 2010).

4.1.2 Extended Example: When Pigs Die

Pig farmers have a strong preference for white pigs and have selectively bred for white coat color since the practice began in medieval Europe (Wiseman, 1986). White coat color in domestic pigs is due to two mutations in the *KIT* gene (Marklund et al., 1998). Although *KIT* mutations control coat color in other mammals, they are often responsible for pigmentation disorders. *KIT* mutations are often lethal in mice, for instance, having a pleiotropic effect on cell development concerning skin, blood, and the small intestines, and may affect hearing, too. The dominant white allele in domestically bred pigs has a stronger effect on pigmentation than it does in mice, but without the selective disadvantage observed in mice. Indeed, there appears to be a selective *advantage* for white domestic pigs over non-white domestic pigs.

Imagine that a laboratory has conducted a study of the selective disadvantage of non-white domestic pigs to white domestic pigs.[6] Coat color in domestic pigs, in our toy theory, is controlled by two alleles at a single locus, the recessive i for color and I allele for the dominant white phenotype. The quantities that interest us include:

1. **Color**: The color of a pig's coat is represented by the binary variable C, which takes the value 1 for white pigs and 0 for non-white pigs.
2. **Genotype**: The discrete variable *Gene* has the value 0 for II, 1 for Ii (or, equivalently, iI), and 2 for ii.
3. **Generation**: The discrete variable G has the value $i - 1$ for the ith generation, for $i \in \{1, 2, 3, 4, 5\}$.
4. **Frequency of white pigs**: $Freq_{C=1}$ gives the proportion of white pigs in a specified set of pigs.

In addition to these quantities, we shall use $N_1, \ldots, N_{15}$ to denote the 15 pens in which a pig can live, $t_1, \ldots, t_{1500}$ as the set of randomly and uniquely assigned identification numbers for all 1500 pigs in the study, and the relation $Lives(x, y)$ which holds when x is a pig, y is a pen, and x lives in y. The study began with a population of 100 pigs and data was collected for five generations. Each generation increased by 100 pigs, and each pen contained 100 pigs of only one generation.

Now let us specify the knowledge base, Γ_δ. In addition to the genetic information just mentioned, we know that in the initial drove of 100 pigs the distribution of the genotypes is: 0.09 II, 0.42 Ii, and 0.49 ii. It follows from this distribution that the frequency of I alleles is 0.3, the frequency of i alleles is 0.7, and the proportion of white pigs is $51/100$. Let's suppose that in any pig population the distribution of genotypes is multinomial, $M\left(p_I^2, 2p_I(1 - p_I), (1 - p_I)^2\right)$, where p_I is the proportion of I alleles. The data covers five generations of random mating starting with our initial drove: p_I is 0.3 for $G = 0$, within [0.348, 0.367] for $G = 1$, within [0.401, 0.439] for $G = 2$, within [0.450, 0.512] $G = 3$, and within [0.502, 0.576] for $G = 4$. We use 'S' to denote the entire population in the study. Also included in Γ_δ are the logical consequences of this information. For example, it follows that p_i when $G = 2$ is between $[.561, .599]$. The following table collects the general information in Γ_δ that we'll use in our example.

In addition to this general knowledge in Γ_δ, there is also specific knowledge about some of the pigs. Pig t_{18} lives in N_{12} and $G(t_{18}) = 1$. Pig t_{1351} lives in some pen or another. Pig t_{662} lives in N_{10}. The set of pigs $\{t_{1428}, t_{1366}, t_{1250}, t_{986}\}$ are all members of the second generation. Pig t_{333} lives in N_6. Pig t_{333} is White.

While it is immediately clear how to represent specific information in logical form, let's pause to consider how general knowledge is represented in Γ_δ. Quantity (4), the frequency of white pigs, implicitly refers to different reference classes of pigs. For instance, the frequency of white pigs in the fourth generation is between 0.642 and 0.825, and between 0.628 and 0.782 in the total population of 1500 pigs in the study. Within EP, each claim is represented by

---

[6] From this point on, our example uses fictitious data and replaces the actual genetic mechanism for coat color with a simplified mechanism.

Table 4.2: General Knowledge in Γ_δ

G	p_I	p_{II}	p_{iI}	$Freq_{C=1}$	$|G|$
0	.3	.09	.42	.51	100
1	[.348, .367]	[.121, .135]	[.441, .479]	[.562, .614]	200
2	[.401, .439]	[.161, .193]	[.450, .526]	[.611, .719]	300
3	[.450, .512]	[.203, .262]	[.439, .563]	[.642, .825]	400
4	[.501, .576]	[.252, .332]	[.426, .574]	[.678, .906]	500
S	[.433, .485]	[.193, .245]	[.436, .538]	[.628, .782]	1500

$\%x(Freq_{C=1}(x), G_{=1}(x), [.642, .825])$ and
$\%x(Freq_{C=1}(x), S(x), [.628, .782])$, respectively.

$G = 1$ and S are reference formulas, denoting reference classes for which we have statistics for the target formula, $Freq_{C=1}$. The fact that $G = 1$ is a proper subset of S is also represented in Γ_δ.

With this setup, we may illustrate some assignments of evidential probability relative to Γ_δ:

1. $Prob(Gene(t_{18}) = 0, \Gamma_\delta) = [0.203, 0.261]$. The frequency of the II genotype in generation 4 $(G = 3)$ is $[0.203, .261]$, but the overall frequency of the II genotype in the full population of the study is $[0.193, 0.245]$, which conflicts with the frequency in generation 4. However, the **Rule of Specificity** advises that we ignore the general statistic for S, $[0.193, 0.245]$, in favor of the statistic for the class $G = 1$, $[0.121, 0.135]$.

2. $Prob(White(t_{18}) = 1, \Gamma_\delta) = [.562, .614]$. Even though the overall frequency of white pigs is $[.628, .782]$, the probability that t_{18} is a white pig is between $[.562, .614]$. Since $\{x | x \in G = 1\} \subsetneq S$, $\%x(Freq_{C=1}(x), G_{=1}(x), [.562, .614])$, and $\%x(Freq_{C=1}(x), S(x), [.628, .782])$ are in Γ_δ and no other information relevant to t_{18} is in Γ_δ, then by **Specificity** we select the more specific reference class statistic.

3. $Prob(White(t_{100}) = 1, \Gamma_\delta) = [.628, .782]$. All that we know about pig t_{100} is that it is in the experiment. Since there is no information in Γ_δ about t_{100} belonging to a subclass which dominates the frequency information for the population, the frequency information for S is used.

4. $Prob(Gene_{=0}(t_{333}), \Gamma_\delta) = [.307, .313]$. We know that t_{333} lives in pen N_6, but there is no information about the proportion of white pigs in N_6. So, although $N_6 \subsetneq S$ and $t_{333} \in N_6$, there is no direct inference statement in Γ_δ recording the proportion of white pigs in N_6. We do know that t_{333} is white, however. Since White pigs are a sub-class for which we have statistics, **Specificity** advises we use the estimate of the proportion of those that are of genotype II, $[.307, .313]$,

rather than the estimate for the proportion of S that are II, $[.193, .245]$.

5. We select 100 pigs from the full population S and observe that 17 are afflicted by polydactyly. If this is all that we know, we may perform a confidence interval analysis and assign a probability of 0.95 that the rate of polydactyly in S is between 12% and 21%. If 0.95 counts as a level of practical certainty, we will simply accept *by default* that between 12% and 21% of the pigs in S have dewclaws.

   Nevertheless, suppose that the rate of polydactyly in domestic pigs is very well understood in the literature to be (A) between 0% and 11% about 80% of the time, (B) between 12% and 21% about 15% of the time, and (C) greater than 21% no more than 5% of the time. In fact, we can see that the observed frequency of 17 in 100 of observed cases (E) given A, B, and C is $.10, .50$, and $.40$, respectively. With this (approximate) distribution of long-run frequencies of polydactyly in domestic swine specified, we can apply a *Bayesian analysis* of our estimate of polydactyly in our population,

$$P(E \mid B) = \frac{P(B \mid E)P(B)}{P(B \mid E)P(B) + P(A \mid E)P(A) + P(C \mid E)P(C)} \approx 0.429.$$

   $P(E|B) \approx 0.429$ conflicts with the interval $[0.95, 1]$ for the hypothesis that between 12% and 21% of the pigs in S have dewclaws, which we constructed by confidence methods. The **Rule of Richness** specifies that the probability based on prior frequency knowledge be preferred to the $[0.95, 1]$ interval, which would *defeat* accepting that between 12% and 21% of the pigs in S have dewclaws.

6. Suppose that a pen is chosen at random and a pig from that pen, t^*, is selected from that pen at random. If this is all that is known, then $\text{Prob}(White(t^*), \Gamma_\delta) = [.601, .871]$ is the probability that a pig is born white. The interval $[.601, .871]$, calculated by taking the average of the probabilities for being white in each generation, does not conflict with the overall proportion of white pigs in the population, $[.628, .782]$. So $[.628, .782]$ is not eliminated in favor of $[.601, .871]$ by the **Rule of Richness**, because the two statistics are not in conflict. Both remain relevant.

   Now suppose that in addition we know two things about t^*. First, t^* is the offspring of t_{1428} and t_{986}, both of which are in $G = 1$. So, $t^* \in G = 2$ and the frequency of white pigs in $G = 2$ is $[.611, .719]$. Second, t^* has the genotype Ii, and suppose that the proportion of pigs of genotype Ii that are white is exactly $.57$. Both $[.57, .57]$ and $[.611, .719]$ are in conflict with each other, and with $[.601, .871]$. But, by **Specificity**, $[.601, .871]$ is ignored. Then by **The Principle of Strength**, $\text{Prob}(White(t^*), \Gamma_\delta) = [.57, .57]$.

4.2 Representation

For the finite set $\mathcal{M}$ of support models M_i ($1 \leq i \leq n$) defined on the pair $\langle \chi, \Gamma_\delta \rangle$, if the proportion of the models in $\mathcal{M}$ of $\varphi_1, \ldots, \varphi_n$ that are also models of χ is between $[l, u]$, then

$$\varphi_1^1, \ldots, \varphi_n^1 \mathrel{|\!\approx} \psi^1,$$

which in $\mathcal{L}^{ep}$ is represented as:

$$\bigwedge_i \%x(\tau(x), \rho(x), [l', u'])_i^1 \bigwedge_j \varphi_j^1 \mathrel{|\!\approx} \langle \chi, [l, u] \rangle^1,$$

where the LHS consists of the accepted evidence in the knowledge base Γ_δ (where $\delta = 0$), i.e., the conjunction of all direct inference statements ($\%x(\tau(x), \rho(x), [l, u])^1$) and all logical knowledge about relationships between classes (φ^1). The entailment relation $\mathrel{|\!\approx}$ is non-monotonic and the RHS is the statement ψ asserting that the target sentence χ is assigned $[l, u]$, where ψ is true when the proportion of support models of

$$\bigwedge_i \%x(\tau(x), \rho(x), [l', u'])_i^1 \bigwedge_j \varphi_j^1$$

that also satisfy χ is between $[l, u]$.

4.3 Interpretation

The EP semantics answer to

$$\varphi_1^{X_1}, \ldots, \varphi_n^{X_n} \mathrel{|\!\approx} \psi^?,$$

is restricted to when $\varphi_1^1, \ldots, \varphi_n^1$, so Y is trivially 1. When ψ expresses $\langle \chi, [l, u] \rangle$, then the assignment of evidential probability *to* χ given $\varphi_1, \ldots, \varphi_n$ is calculated by finding the proportion of the models in $\mathcal{M}$ of $\varphi_1^1, \ldots, \varphi_n^1$ that are also models of χ.

The relation $\mathrel{|\!\approx}$ is not governed by probabilities assigned to sentences in Γ_δ. Instead, the meaning of $\mathrel{|\!\approx}$ is given by rules for resolving conflicts among accepted evidence about frequencies that are relevant to χ. Expressing this behavior of EP in terms of Schema (1.1) allows us to represent several sharp differences between evidential probability and the other approaches in this book.

But Schema (1.1) also allows us to pose two distinct inferential questions for evidential probability. The first question is the question of how to assign evidential probability to a statement given some evidence, which is the Kyburg-Teng semantics for EP. We call this *first-order evidential probability*. The second question concerns the risk associated with inferring an EP probability for a statement on some particular set of evidence. We call this *second-order evidential probability*. To connect second-order EP with first-order EP we need to consider the possible impact false evidence would have on a particular assignment of evidential probability. We call this *counter-factual evidential probability*. In the remainder we show how to supply Schema (1.1) with first-order EP-semantics, present a theory of counter-factual evidential probability, then present a theory of second-order EP and show how to supply Schema (1.1) with second-order EP semantics.

4.3.1 First-order Evidential Probability

To understand the entailment relation in EP it is necessary to decouple the semantics for $\mathrel{\mspace{1mu}\vrule\mspace{1mu}\sim}$ from probabilities. We achieve this by assigning 1 to each sentence in Γ_δ while simultaneously observing that $\mathrel{\mspace{1mu}\vrule\mspace{1mu}\sim}$ does not satisfy the KLM System P semantics (Kraus et al., 1990a) as might otherwise be expected (Kyburg et al., 2007).[7]

Proposition 3 (Properties of EP Entailment). *Let $\models$ denote classical consequence and let $\equiv$ denote classical logical equivalence. Whenever $\mu \wedge \xi, \nu \wedge \xi$ are sentences of $\mathscr{L}^{ep}$,*

Right Weakening: *if $\mu \mathrel{\mspace{1mu}\vrule\mspace{1mu}\sim} \nu$ and $\nu \models \xi$ then $\mu \mathrel{\mspace{1mu}\vrule\mspace{1mu}\sim} \xi$.*
Left Classical Equivalence: *if $\mu \mathrel{\mspace{1mu}\vrule\mspace{1mu}\sim} \nu$ and $\mu \equiv \xi$ then $\xi \mathrel{\mspace{1mu}\vrule\mspace{1mu}\sim} \nu$.*
(KTW) Cautious Monotony: *if $\mu \models \nu$ and $\mu \mathrel{\mspace{1mu}\vrule\mspace{1mu}\sim} \xi$ then $\mu \wedge \xi \mathrel{\mspace{1mu}\vrule\mspace{1mu}\sim} \nu$.*
(KTW) Conclusion Conjunction: *if $\mu \models \nu$ and $\mu \mathrel{\mspace{1mu}\vrule\mspace{1mu}\sim} \xi$ then $\mu \mathrel{\mspace{1mu}\vrule\mspace{1mu}\sim} \nu \wedge \xi$.*

This approach to defining EP-entailment presents challenges in handling disjunction in the premises since the KLM disjunction property admits a novel reversal effect similar to, but distinct from, Simpson's paradox (Kyburg et al., 2007; Wheeler, 2007). This raises a question over how to axiomatize EP. One approach, which is followed by (Hawthorne and Makinson, 2007) and considered in (Kyburg et al., 2007), is to replace Boolean disjunction by 'exclusive-or'. While Hawthorne and Makinson's system O ensures some nice properties for $\mathrel{\mspace{1mu}\vrule\mspace{1mu}\sim}$, it cannot be given a finite axiomatization in terms of Horn rules that is complete (Paris and Simmonds, 2009), and introduces a dubious connective into the object language that is neither associative nor compositional.[8] Another approach explored in (Kyburg et al., 2007) is a weakened disjunction axiom that, together with the axioms of Proposition 3, yield a sub-System P non-monotonic logic that preserves the standard properties of the positive Boolean connectives. This system is incomplete as well, however *qualitative* representations for fragments of EP have been proposed in terms of the classical modal logic EMN (Kyburg and Teng, 2002) and description logic (Swift and Wheeler, 2011) which are complete.

---

[7] The KTW properties (Kyburg et al., 2007) are similar to, but strictly weaker than, the properties of the class of cumulative consequence relations specified by System P (Kraus et al., 1990a). To yield the axioms of System P, replace $(\mu \models \nu)$ by $(\mu \mathrel{\mspace{1mu}\vrule\mspace{1mu}\sim} \nu)$ in **Cautious Monotony** and **Conclusion Conjunction**, and add **Premise Disjunction**: if $\mu \mathrel{\mspace{1mu}\vrule\mspace{1mu}\sim} \nu$ and $\xi \mathrel{\mspace{1mu}\vrule\mspace{1mu}\sim} \nu$ then $\mu \vee \xi \mathrel{\mspace{1mu}\vrule\mspace{1mu}\sim} \nu$.

[8] Example: 'A xor B xor C' is true if A, B, C are; and '(A xor B) xor C' is not equivalent to 'A xor (B xor C)' when A is false but B and C both true.

4.3.2 Counterfactual Evidential Probability

Evidential probability assigns probability to a sentence χ given the set Γ_δ of accepted statements, whereas *counterfactual evidential probability* assigns probability to χ given that some sentence ϕ in Γ_δ is false. Counterfactual EP then is the evidential probability assigned to χ given the replacement of ϕ in Γ_δ by $\neg\phi$.

This conception of counterfactual EP allows you to compare a first-order evidential probability assignment to all possible alterations of an evidence set by calculating first-order EP probability for *all* possible combination of truth values assigned to formulas in Γ_δ. Altering truth assignments to some elements may be relevant to χ (i.e., when $[l,u] \neq [l',u']$) or irrelevant to χ (i.e., when $[l,u] = [l',u']$). The minimal set $\Delta = \{\varphi_1, \ldots, \varphi_m\}$ of sentences in Γ_δ such that *some* combination of unnegated or negated evidential propositions (i.e., $\{\varphi_1^{e_1}, \ldots, \varphi_m^{e_m}\}$ for some $(e_1, \ldots, e_n) \in \{0,1\}^n$) is relevant to χ is called the set of *possibly relevant evidence* for χ with respect to Γ_δ. On this view, the counter-factual evidential probability of χ given $\{\varphi_1^{e_1}, \ldots, \varphi_m^{e_m}\}$ is just what the evidential probability of χ given Γ_δ would be, *were* Γ_δ to be $\{\varphi_1^{e_1}, \ldots, \varphi_m^{e_m}\}$.

The possibly relevant evidence $\Delta \subseteq \Gamma_\delta$ for χ is a set of 2^m evidential probabilities, $\mathrm{Prob}\left(\chi | \varphi_1^{e_1}, \ldots, \varphi_m^{e_m}\right)$, where

1. $e_1, \ldots, e_m \in \{1,0\}$, and
2. $\mathrm{Prob}\left(\chi | \varphi_1^{e_1}, \ldots, \varphi_m^{e_m}\right) = \frac{\|[l,u] \cap [l',u']\|}{\|[l,u]\|}$, when $e_i = 1 \vee e_i = 0$, for all $1 \leq i \leq m$.
3. $\mathrm{Prob}\left(\chi | \varphi_1^{e_1}, \ldots, \varphi_m^{e_m}\right) = [0,1]$ when $e_i = 0$, for all $1 \leq i \leq m$.

Condition (1) specifies that e_i is the truth assignment to φ_i and condition (2) measures the overlap between the actual evidence and an item of possible evidence from a counterfactual argument. Comparing the actual, first-order evidential probability to itself is identity, i.e., 1, which is the special case of (2) when $e_i = 1$, for all $1 \leq i \leq m$. A measure of 0 denotes no overlap between actual and counterfactual evidence.

4.3.3 Second-Order Evidential Probability

Return now to the introduction of this section and to our brief discussion of accepted evidence. Γ_δ is the set of accepted but *uncertain* premises, and we have represented that feature in our semantics for first-order EP entailment by assigning 1 to each premise and supplying a sub-System P entailment relation, $\models$. There is considerable controversy over whether the Kyburgian theory of acceptance should be replaced by a probabilistic account (Jeffrey, 1956; Kyburg, 1961; Carnap, 1968; Levi, 1977; Harper, 1981; Kyburg, 2003, 2007; Levi, 2007; Seidenfeld, 2007). Indeed, the controversy over this issue mirrors another in statistics regarding R. A. Fisher's fiducial argument and the availability of fiducial probabilities for inference, which is a topic we return to in Section §5.

For the purposes of this book we remain neutral on Kyburg's theory of acceptance and instead consider what EP might look like were its first-order logic of probabilities embedded within a 'second-order' probabilistic logic. A theory of second-order EP inference would aim to provide a means to evaluate a sequence of EP-inferences. Rather than assign 1 to each φ_i, we may substitute X_i by the practical certainty interval $[1 - \delta, 1]$ if φ_i is a direct inference statement, and by 1 otherwise. The question is then: how certain should one be that the first-order EP interval attaching to χ accurately bounds the probability of χ, given that the evidential propositions are in fact uncertain to some small degree δ? The apparatus of second-order EP can answer this question. It takes first-order EP as a starting point, by assuming that if $\delta = 0$ then the first-order evidential probability does accurately bound the probability of χ. But for $\delta > 0$, second-order evidential probability is intended to assign an estimate to the 'degree of risk' associated with inferring an EP probability for χ on the basis of Γ_δ. Second-order EP then may be thought to provide a measure of robustness in the evidence for assigning $[l, u]$ to χ, thus providing a measure of confidence in $\text{Prob}(\chi, \Gamma_\delta)$.

A candidate semantics for second-order EP that answers $\varphi_1^{X_1}, \ldots, \varphi_n^{X_n} \approx \psi$? is given by

$$P(\psi) = \sum_{e_1,\ldots,e_m=0}^{1} P\left(\psi | \varphi_1^{e_1}, \ldots, \varphi_m^{e_m}\right) \cdot P\left(\varphi_1^{e_1}, \ldots \varphi_m^{e_m}\right),$$

where ψ states that $P(\chi) \in [l, u]$ (which may in turn be inferred from a first-order EP statement $\text{Prob}(\chi, \Gamma_\delta) = [l, u]$), where each φ_i is either a direct inference statement or logical formula in Γ_δ and where X_i is $[1 - \delta, 1]$ or 1, respectively. The formula χ is just a sentence of $\mathscr{L}^{ep}$, but ψ is the proposition expressing the assignment of $[l, u]$ to χ.

Note that we take the risk intervals $[1 - \delta, 1]$ to be probabilities attached to elements in Γ_δ, and this assumption is critical to give a probabilistic semantics to second-order EP inference. Since there is this connection between second-order EP and probability functions—absent in the case of first-order EP—we do now have a genuine probabilistic logic: $\varphi_1^{X_1}, \ldots, \varphi_n^{X_n} \approx \psi^Y$ if and only if $P(\psi) \in Y$ for all probability functions P satisfying $P(\varphi_1) \in X_1, \ldots, P(\varphi_n) \in X_n$.

We shall make two further simplifying assumptions in this book. The first is that probability is distributed uniformly across an EP interval. This allows us to set

$$P\left(P(\chi) \in [l, u] | \varphi_1^{e_1}, \ldots, \varphi_m^{e_m}\right) = \frac{|[l, u] \cap [l', u']|}{|[l', u']|}$$

where $[l', u']$ is the counterfactual evidential probability for χ given $\{\varphi_1^{e_1}, \ldots, \varphi_m^{e_m}\}$, i.e., $\text{Prob}\left(\chi, \{\varphi_1^{e_1}, \ldots, \varphi_m^{e_m}\}\right) = [l', u']$. Note that we can restrict our attention to possibly relevant evidence for χ:

$$P\left(P(\chi) \in [l, u] | \varphi_1^{e_1}, \ldots, \varphi_m^{e_m}\right) = P\left(P(\chi) \in [l, u] | \varphi_{i_1}^{e_{i_1}}, \ldots, \varphi_{i_l}^{e_{i_l}}\right)$$

Where $\left\{ \varphi_{i_1}^{e_{i_1}}, \ldots, \varphi_{i_l}^{e_{i_l}} \right\} \subseteq \left\{ \varphi_1^{e_1}, \ldots, \varphi_m^{e_m} \right\}$ is the subset of possibly relevant evidence (§4.3.2).

Secondly we assume, in the absence of evidence otherwise, that items of evidence are probabilistically independent. Hence

$$P\left(\varphi_1^{e_1}, \ldots \varphi_m^{e_m}\right) = P\left(\varphi_1^{e_1}\right) \ldots P\left(\varphi_m^{e_m}\right).$$

Chapter 5
Statistical Inference

An important application of probability theory is the use of statistics in science, in particular classical statistics as devised by Fisher and Neyman and Pearson. Good introductions to this type of statistics are provided in (Barnett, 1999) and in (Mood et al., 1974). We should emphasize that classical statistics is not an uncontroversial tool for reasoning statistically, and that it is sometimes in direct disagreement with the other major theory of statistical inference treated in this book, Bayesian statistics. A good and accessible overview of the problems that beset the classical statistical account of statistics is given by (Howson and Urbach, 1993). But these problems do not directly concern us here. In this section we merely indicate when and how it can be accommodated by Schema (1.1).

5.1 Background

This section briefly discusses the inferential status of classical statistics, and then considers two attempts at providing an inferential representation of it, namely by fiducial probability and by evidential probability.

5.1.1 Classical Statistics as Inference?

Classical statistical procedures concern probability assignments over data D relative to a statistical hypothesis H, namely $P_H(D)$. The crucial property of these procedures is that they only involve direct inference: they involve the derivation of the probability of some event for a given statistical hypothesis. In turn, the probability assignments feature in other functions defined over the sample space, such as estimators and tests. In this book we restrict attention to Neyman-Pearson tests and Fisher's theory of estimation.

R. Haenni et al., *Probabilistic Logics and Probabilistic Networks*, Synthese Library 350,   49
DOI 10.1007/978-94-007-0008-6_5, © Springer Science+Business Media B.V. 2011

Classical Statistical Procedures.

Let Ω_H denote a statistical model, consisting of statistical hypotheses H_j with $0 \leq j < n$ that are mutually exclusive. The assignment that H_j is true is written h_j. Let Ω_D be the sample space, consisting of observations of binary variables D_i. For a specific sample, i.e. an assignment to a set of D_i, we write d_s^e, where s is an m-vector of indices i of the variables D_i, and e is a binary m-vector encoding whether D_i is true or false, so that $d_s^e = d_{s(1)}^{e(1)} \cdots d_{s(m)}^{e(m)}$. If s is simply $\langle 1, 2, \ldots, m \rangle$ we write d^e, and if s and e are not given by the context, we write d for a specific sample and D for data when treated as a variable. All these assignments are associated with subsets of the sample space, $D \in \Omega_D$.

Restricting Ω_H to $n = 2$, we can compare the hypotheses h_0 and h_1 by means of a Neyman-Pearson test function. See Barnett (1999) and Neyman and Pearson (1967) for the details.

Definition 11 (Neyman-Pearson Hypothesis Test). Let T be a set function over the sample space Ω_D,

$$T(D) = \begin{cases} 1 & \text{if } \frac{P_{h_1}(D)}{P_{h_0}(D)} > t, \\ 0 & \text{otherwise,} \end{cases} \tag{5.1}$$

where P_{h_j} is the probability over the sample space determined by the statistical hypothesis h_j. If $T = 1$ we decide to reject the null hypothesis h_0, else we reject the alternative h_1.

The decision to reject is associated with a significance and a power of the test:

$$\text{Significance}_T(D) = \int_{\Omega_D} T(D) P_{h_0}(D) d\omega,$$

$$\text{Power}_T(D) = \int_{\Omega_D} T(D) P_{h_1}(D) d\omega.$$

In their fundamental lemma, Neyman and Pearson prove that the decision has optimal significance and power for, and only for, likelihood-ratio test functions T of Equation (5.1).

Among a larger set of statistical hypotheses, taking $n \geq 2$, we may also choose the best performing one according to an estimation procedure. The maximum likelihood estimator advocated by Fisher varies with the probability that the hypotheses assign to points in the sample space. See Barnett (1999) and Fisher (1956) for the details.

Definition 12 (Maximum Likelihood Estimation). Let Ω_H be a model with hypotheses H_θ, where $\theta \in [0, 1]$. Then the maximum likelihood estimator is

$$\hat{\theta}(D) = \left\{ \theta : \forall h_\theta' \left(P_{h_{\theta'}}(D) \leq P_{h_{\theta'}}(D) \right) \right\}, \tag{5.2}$$

where D is again generic data. So the estimator is a set, typically a singleton, of those values of θ for which the likelihood of h_θ is maximal. The associated best hypothesis we denote with $h_{\hat\theta}$.

We may further compute the so-called confidence interval. Usually parameter values that are near to the estimator assign only slightly smaller probabilities to the sample, so that we can for example define a region R of parameter values for which the data are not highly unlikely, $R(D) = \{\theta : P_{h_\theta}(D) > 1\%\}$. The definition of the region of parameter values that is usually termed the symmetric 95% confidence interval follows this idea but is slightly more complicated:

$$C_{95}(D) = \left\{ \theta : |\theta - \hat\theta| < \lambda \text{ , and } \int_{\hat\theta - \lambda}^{\hat\theta + \lambda} P_{H_\theta}(D)d\theta = .95 \right\}.$$

In this way every element of the sample space D is assigned a region of parameter values, which expresses the quality of the estimate.[1]

Procedure vs Inference.

These are just two examples of classical statistical procedures. More generally, running a classical procedure comes down to collecting specific data d, going to the corresponding set or point d in the sample space, and then looking up the values of functions like T, $\hat\theta$, and C_{95} for that point. Typically, these values are themselves associated with certain probability assignments over the sample space. For example, the Neyman-Pearson test function may indicate that the null hypothesis can be rejected, but with that indication comes a probability for error, the significance level. And the Fisher estimation function may be the value for the mean of some normal distribution, but there is always some confidence interval around the mean. Much of classical statistics is concerned with these probabilities associated with the tests and estimations.

    Classical statistical procedures make constant use of direct inference, which concerns probability assignments over the sample space Ω_D. Moreover, the procedures are about the hypotheses H_θ. They concern the rejection or a best estimate for hypotheses, and they associate probabilities of error to these decisions concerning hypotheses. Most importantly, the procedures do not assign probabilities to the space Ω_H. Neyman and Pearson (1967) are emphatic that classical statistical procedures must not be seen as inferences about hypotheses to start with. They guide decisions

---

[1] The confidence region is sometimes interpreted inferentially, as if it expresses a probability for the parameter value to lie within the designated interval. Strictly speaking this is not correct. Rather it is the region within which the estimator $\hat\theta$ will fall 95% of the time if we were to repeat our estimations, under the assumption that $\hat\theta$ is the true parameter value. Since $\hat\theta$ is itself an estimation, we might say that $C_{95}(D)$ is an estimation of the region that contains $\hat\theta$ 95% of the time. For an inferential reading of confidence intervals, we need to make resort to a fiducial argument; see section §5.1.2.

about them, they have certain error rates associated with them, but they are not inferential.

Because the decisions are not cast in terms of probabilities over hypotheses, it is doubtful whether we can faithfully accommodate classical statistics in the inferential problem of Schema (1.1). In representing the procedures involving errors of estimations and tests, we cannot employ probabilities assigned to hypotheses: function values such as $T(D)$, $\hat{\theta}(D)$, and $C_{95}(D)$ cannot be expressed in a probability assignment over hypotheses, or anything like it. There are, in other words, objections to forcing classical statistics into the inferential format, and so there are objections to formulating a probabilistic logic for it. Nevertheless, in the following we investigate how strict these objections are, and what an inferential format may look like.

5.1.2 Fiducial Probability

One way of capturing classical statistics in terms of inferences has been suggested by Fisher (1930, 1935, 1956). Applications of his so-called fiducial argument allow us to derive probability assignments over hypotheses from a classical statistical procedure, so that classical statistics may after all be reconciled with an inferential attitude.

The Fiducial Argument.

Suppose that it is known that a quantity, F, is distributed normally with an unknown mean, μ, and a known variance σ^2 of 1. After drawing a sample of size 1, it is observed of that sample that F takes the (continuous) value r. Since $\mu - r$ is known to be normally distributed, we can look up the table value of the probability that $|\mu - r|$ exceeds any given amount. For instance, if $r = 10$ is observed, we can, by direct inference, infer that the probability that μ is between 9 and 11 is 0.68. This example is an illustration of Fisher's fiducial inference. The argument draws an inference from observable data (that F takes r in the sample) to a statistical hypothesis (the mean μ of F) without following the form of inverse inference: while the inference relies upon the statistical knowledge concerning the distribution of F, it does not appear to rely upon any knowledge concerning the distribution of μ. A probability assignment over μ can thus be derived without assuming any prior probability assignment.

The fiducial argument is controversial, however, and its exact formulation is a subject of debate. The controversy stems from Fisher setting out goals for fiducial inference that do not appear mutually consistent (Seidenfeld, 1992), and is compounded by Fisher's informal account of fiducial probability. Bayesian statisticians such as Lindley (1958) and Good (1965) argue that fiducial probability is based on an implicit, 'improper' prior probability, in part because Fisher himself thought that the output of a fiducial argument should always be available for Bayesian inference. But if Lindley and Good are right, this would bring us back to the worry that classical statistics cannot be done justice in an inferential schema.

Dempster (1963) and Hacking (1965) show that the fiducial argument relies on specific assumptions concerning statistical relevance, and Fisher's own informal description of how to resolve conflicting statistical laws has suggested to some that there is no principled theory of fiducial inference at bottom. We return to this point in the next section when we discuss evidential probability.

Dawid and Stone (1982) provide a general characterisation of the set of statistical problems that allow for application of the fiducial argument, using so-called functional models. As will be spelled out below, the functional relation between data, hypotheses, and probabilistic elements must be smoothly invertible. But many statistical problems fall outside this set, pointing to the inconsistency we mentioned at the start. If one insists that the output of the fiducial argument should be available for Bayesian inference, then it seems rather ad hoc to say that the inferential schema for classical statistics is supposed to apply only to certain cases. On the other hand, the point can be turned around to highlight when it is appropriate to presume knowledge of a full, joint distribution.

Deriving Fiducial Probability from Functional Models.

Functional models reveal the limits of the fiducial argument, but as Kohlas and Monney (2008) show, they also provide the starting point for an adapted and more general version of the fiducial argument, based on what they call assumption-based reasoning. Their view is not entirely new: Arthur Burks (1953) was an early proponent of a presupposition based view of inductive inference.

New in the approach of Kohlas and Monney is that they employ assumption-based reasoning to generate degrees of support and possibility. The theory of support and possibility has already been discussed in §3. Its use in the adapted version of the fiducial argument is that the evidence is not taken to induce a fiducial probability over the statistical hypotheses. Rather it induces degrees of support and probability; these degrees coincide in a probability measure only if the functional model is of the type that makes the traditional fiducial argument applicable.

Let the model Ω_H and sample space Ω_D be as in the foregoing. A functional model can be characterised as follows.

Definition 13 (Functional Model). Let Ω_H be a statistical model, and Ω_W a set of stochastic elements ω. A functional model consists of a function

$$f(H_\theta, \omega) = D, \qquad\qquad (5.3)$$

and a probability assignment $P(\omega)$ over the stochastic elements in Ω_W.

That is, a functional model relates every combination of a statistical hypothesis H_θ and an assumed stochastic element ω to data D. These stochastic elements are elements of the space Ω_W. They must not be confused with the d^e that denote valuations of the variables D, which will in the following be abbreviated with d. We define $V_d(h_\theta)$ as the set of ω for which $f(h_\theta, \omega) = d$, where d represents an assignment to the variable D. We can then derive the likelihoods of the hypotheses h_θ:

$$P(d|h_\theta) = \int\limits_{\omega \in V_d(h_\theta)} P(\omega)d\omega. \tag{5.4}$$

A functional model is smoothly invertible if there is a function $f^{-1}(H_\theta, D) = \omega$, so that relative to the sample D, each hypothesis H_θ is associated with exactly one stochastic element ω.

Now consider the hypothesis $h_I = \cup_{\theta \in I} h_\theta$, determined by an interval I. The set $U_d(\omega) = \{h_\theta : f(h_\theta, \omega) = d\}$ covers exactly those hypotheses that point to the sample d under the assumption of the stochastic element ω. We define the support Sup and the possibility Pos of the data d for the hypothesis h_I as

$$Sup_{h_I}(d) = \{\omega : U_d(\omega) \subset h_I\} \tag{5.5}$$

$$Pos_{h_I}(d) = \{\omega : U_d(\omega) \cap h_I \neq \emptyset\}. \tag{5.6}$$

The set $Sup_{h_I}(d) \subset \Omega_W$ consists of all stochastic elements that, together with the data d, entail the hypothesis h_I, while the set $Pos_{h_I}(d) \subset \Omega_W$ consists of all those stochastic elements that, together with the data d, leave the hypothesis h_I possible. In the terminology of section §3 the former are the arguments for h_I, while the latter are the complement of the arguments against h_I.

We can now define the degrees of support and possibility for h_I accordingly, as

$$dsp(h_I) = \int\limits_{\omega \in Sup_{h_I}(d)} P(\omega)d\omega, \tag{5.7}$$

$$dps(h_I) = \int\limits_{\omega \in Pos_{h_I}(d)} P(\omega)d\omega. \tag{5.8}$$

So the degrees of support and possibility derive from the probability assignment over Ω_W, the stochastic elements in the functional model. As illustrated in Figure 5.1, the probability over Ω_W is transferred to degrees of support and possibility over the model Ω_H.

The standard fiducial argument is a special case of the derivation of fiducial degrees of support and possibility. If f is smoothly invertible, the sample d entails a one-to-one mapping between Ω_W and Ω_H. In Figure 5.1, the area d must then be replaced by a line. As a result, the degrees of support and possibility for h_I collapse to a single sharp value, $P_d(h_I) = dsp(h_I) = dps(h_I)$. The probability distribution over ω can be transferred directly onto the single hypotheses h_θ.

The generalised fiducial argument is nice from the perspective of the progicnet programme, because it enables us to incorporate classical estimation in the inferential format of Schema (1.1), even if we do not have a smoothly invertible functional relation. But as indicated in the foregoing, the standard fiducial argument does as-

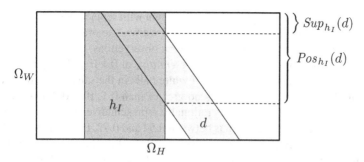

Fig. 5.1: The rectangle shows all combinations of H_θ and ω. Those combinations for which $f(H_\theta, \omega) = d$ are included in the shaded area. The hypothesis h_I, which spreads out over a number of h_θ, has a minimal and a maximal probability: it is at least as probable as the union of stochastic elements ω that entail d with every element $h_\theta \in h_I$, and it is at most as probable as the union of stochastic elements ω that entail d with at least one element $h_\theta \in h_I$.

sume the invertibility of f. For the sake of simplicity, we will only return to this more narrow application of functional models for fiducial probability in section §12.

5.1.3 Evidential Probability and Direct Inference

Direct inference, which is sometimes called the *frequency principle*, assigns a probability to a particular individual or event having a property on the basis of known statistical frequencies of that property occurring. Sometimes we want to assign a probability to a statistical hypothesis on the basis of observed data, however, and we remarked above that the fiducial argument was Fisher's strategy to reduce 'indirect' inference to a form of direct inference.

For Fisher, the applicability of a statistical regularity R to a particular individual t requires the specific knowledge that R is relevant to t and total ignorance of any competing law R' that is also relevant to t. In simplest terms, if all we know of t is that the statistical regularities of R apply to t, then we may treat t as a random event with respect to R and infer that those regularities characterize t. But there are complications.

One problem is that often there are several statistical regularities that we know apply to t, but this collection may yield conflicting assignments of probability. Another complication concerns precisely how ignorance and knowledge is supposed to work to effect randomization in a successful inference. We've seen a theory that addresses the issue of conflict in §4. Indeed, evidential probability fully embraces the view that there are forms of statistical inference after all (cf. Neyman (1957)). Now we focus on the issue of epistemic randomness.

Suppose we observe a sample of n balls, drawn with replacement, from an urn of black and white balls of unknown number and we wish to draw an inference about the proportion of white balls in U from those n observations.

A sample of n draws is *rationally representative* at 0.1 if and only if the difference between the observed proportion of white balls in the sample n and the actual proportion of white balls in the urn is no greater than 0.1. If 100 draws are made, 62 of which are white, the sample is rationally representative at 0.1 just in case the proportion of white balls in the urn is between 0.52 and 0.72. If we know that the urn is composed of exactly as many white balls as black, then the sample n would *not* be rationally representative at 0.1, for we would have then expected the proportion of white balls to be between 0.4 and 0.6.

Suppose we do not know the proportion of white balls and black balls, which is to suppose that we wish to estimate the proportion of white to black in the urn from observing the proportion of samples from the urn. We might, however, know enough about our experimental procedure to accept that it is rationally representative. In such cases there are properties of rationally representativeness that may be exploited to make informative inferences. By using a normal distribution approximation for the binomial frequencies, we may infer that between 95% and 100% of all samples of n-fold sequences ($n = 100$) of independent outcomes from a binomial process are rationally representative at 0.1. Assuming that our sample of 100 balls is a random member of the class of all 100-fold trials, we may then infer non-trivial conclusions about the population. Specifically, we may infer that the evidential probability that the sample of 62 white balls of 100 is rationally representative at 0.1 is $[0.95, 1]$. In other words, the evidential probability is no less than 0.95 that the proportion of white balls in the urn is between 0.52 and 0.72. If our threshold for rational acceptance is 0.95 or less, then we would simply say that it is rational to accept that the proportion of white balls is between 0.52 and 0.72.

But what grounds do we have for making the substantive assumption that our sample of 100 balls *is* a random member of the collection? This question concerns where to place the burden of proof, which concerns the interplay between specified knowledge and specified ignorance in statistical estimation. The policy of evidential probability is to treat randomness as a default assumption (Kyburg and Teng, 1999; Wheeler, 2004; Wheeler and Pereira, 2004) rather than an explicit condition that is satisfied. Randomness is assumed to hold on the basis of passing tests designed to detect bias, which is a policy matched by practice. Asserting that a sample passes such tests does not entail that the sample is random, but rather indicates that no evidence of bias was found in light of this battery of tests. That a sample 'satisfies' epistemic randomness, then, translates to the absence of evidence that the sample is biased. Some may be uncomfortable with assuming that epistemic randomness holds without direct evidence, but notice that if we could observe the rational representativeness property directly then we would not be in the position of needing to perform an inverse inference, for having direct evidence that a sample is representative is in effect is to know the statistical distribution of the population already.

There is a price to viewing the fiducial argument in terms of a non-demonstrative logic of probabilities rather than a species of probabilistic logic. EP yields probability assignments that cannot be readily used in subsequent inference (Wheeler and Williamson, 2010), assignments that are often incompatible with Bayesian methods (Kyburg, 2007; Seidenfeld, 2007), and yield credal probabilities that cannot (always) be managed by conditionalization (Kyburg, 1974; Levi, 1977). Even so, the approach offers rich insight into understanding Fisher's fiducial argument in particular, and classical statistical inference *qua* inference in general.

5.2 Representation

We now consider, both for inferences concerning fiducial and evidential probability, how the inferences may be represented in Schema (1.1).

5.2.1 Fiducial Probability

Using functional models, we can represent classical statistical estimation in Schema (1.1) by employing fiducial degrees of support and possibility:

$$(\{(h_\theta, \omega) : f(h_\theta, \omega) = d\})^1, \omega^{P(\omega)} \models h_I^{[dsp, dps]}. \tag{5.9}$$

Here h_I refers to the statistical hypothesis of interest, as defined above, and ω to the stochastic element of the functional model, over which the probability assignment $P(\omega)$ is given. The function f determines how these elements relate to possible data D. The set $\{(H_\theta, \omega) : f(H_\theta, \omega) = d\}$ refers to all combinations of hypothesis and stochastic element that lead to d as evidence. The set of these combinations (H_θ, ω) is given probability 1, meaning that the evidence occurred. This induces an interval-valued assignment over the hypothesis h_I.[2]

As said, in the case that the function f is smoothly invertible the interval-valued assignment to h_I collapses to a sharp probability value. The fiducial argument then becomes:

$$(\{(H_\theta, \omega) : f(H_\theta, \omega) = d\})^1, \omega^{P(\omega)} \models h_I^{P_d(h_I)}, \tag{5.10}$$

where $P_d(h_I) = Sup_{h_I}(d) = Pos_{h_I}(d)$. In the second part of this book we will mainly explore the use of networks in determining sharp fiducial probabilities.

---

[2] Clearly, in quite a few cases the interval attached to the hypothesis will be $[0, 1]$, but it is beyond the scope of this book to investigate when this occurs.

5.2.2 *Evidential Probability and the Fiducial Argument*

Functional models offer one way of representing Fisher's fiducial argument by transforming statistical reduction into a form of demonstrative inference. But Fisher was very explicit in arguing that statistical reduction should be viewed as a type of logical, *non-demonstrative* inference (Fisher, 1922, 1936). Unlike demonstrative inference from true premises, the 'validity' of a non-demonstrative, uncertain inference can be undermined by additional premises: a conclusion may be drawn from premises supported by the total evidence available now, but new premises may be added that remove any and all support for that conclusion. Fisher, it turns out, was an early proponent of non-monotonic logic.

First-Order EP and the Fiducial Argument.

With this in mind we can represent the fiducial argument as a special case of a first-order EP argument by

$$(\bigwedge_i \%x(\tau(x), \rho(x), [l', u'])_i^1 : f(\chi, \rho(x)_{[x|\omega x]}) \geq \lambda\}) \bigwedge_j \varphi_j^1 \models \langle \chi, [l, u] \rangle^1. \quad (5.11)$$

Here $f(\chi, \rho(x)_{[x|\omega]}) = \lambda$ selects a subset of relevant statistics for χ from a knowledge base Γ_δ by a substitution of variable x by constants ω induced by χ such that each relevant statistical statement is rationally representative of χ to degree λ. The function f, then, effects the reduction of 'indirect' inference to a form of direct inference. Then the machinery of EP (§4.1.1) is run on this restricted set of rationally representative direct inference statements to assign the probability interval $[l, u]$ to hypothesis χ.

Second-Order EP and the Fiducial Argument.

If the fiducial argument is viewed as a restricted form of first-order EP inference, then it is natural to inquire about the robustness of a particular assignment of evidential probability made by a fiducial argument. Then

$$(\bigwedge_i \%x(\tau(x), \rho(x), [l', u'])_i^{X_i} : f(\chi, \rho(x)_{[x|\omega x]}) \geq \lambda\}) \bigwedge_j \varphi_j^{X_j} \models \langle \chi, [l, u] \rangle^Y,$$

represents the second-order probability Y to the statement expressing that the first-order EP probability of χ is $[l, u]$, where the LHS includes the logical and possible, relevant statistical evidence under the 'rational representativeness to degree λ' restriction, $f(\chi, \rho(x)_{[x/\omega x]}) \geq \lambda$. The difference between second-order EP probability and a second-order EP probability of a fiducial argument is that in the latter case the premises that (partially) determine the function f are held fixed rather than allowed to vary. Thus, the fiducial argument here is a type of conditional second-order EP probability. We return to this briefly in §5.3.2.

5.3 Interpretation

We show how the inferences represented in Schema (1.1) can be interpreted as concerning inferences of fiducial and evidential probability, respectively.

5.3.1 Fiducial Probability

Insofar as classical statistical procedures can be viewed as inferences, we can also interpret the central question of Schema (1.1) as an inference in classical statistics. The set of probability distributions that we start out with is defined over the space of hypotheses and stochastic elements, $\Omega_H \times \Omega_W$. One type of premise in the inference is then given by the data d, represented as a set of combinations of hypotheses H_θ and stochastic elements ω, namely $\{(h_\theta, \omega) : f(h_\theta, \omega) = d\}$. The premise given by the data is that we restrict the set of probability distributions to those that assign $P(\{(h_\theta, \omega) : f(h_\theta, \omega) = d\}) = 1$. Another type of premise is presented by the probability distribution over the stochastic elements ω: we restrict the set of probability functions to those for which the marginal $P(\omega)d\omega$ is given by some specific function. With these two premises in place, we can derive the degrees of support and possibility using the general framework of this book.

We must emphasise, however, that this inferential representation, and interpretation, of classical statistical procedures runs into difficulty with dynamic coherence. The foregoing shows that for each particular sample d we can set up a generalised fiducial argument and derive degrees of support and possibility. However, as shown by (Seidenfeld, 1992), additional samples d' cannot be incorparated in the inference by a simple conditionalisation on that data. That is, if we find additional data sets d' and compute the degrees of support and possibility as $dsp(h_I|d')$ and $dps(h_I|d')$, using the fiducial degrees of support and possibility that were based on d as priors, then the results may deviate from the result of running the fiducial argument with $d \wedge d'$ and deriving degrees of support and possibility directly, using a different functional relation. As is explained further by Seidenfeld, this reveals that fiducial probabilities are sensitive to a typically Bayesian problem, namely that the principle of indifference does not single out a unique prior. This severely restricts the use of the generalised fiducial argument in a logical representation of classical statistical inference.

5.3.2 Evidential Probability

First-Order EP and the Fiducial Argument.

To interpret

$$\varphi_1^{X_1}, \ldots, \varphi_n^{X_n} \approx \psi^?,$$

as an EP reconstruction of fiducial inference, we first replace $\approx$ by $\vDash$ of first-order EP-entailment, since the fiducial argument is interpreted on this view to be a non-demonstrative inference. Second, we interpret the premises to include logical information and direct inference statements, just as in §4, and add to this a function $f(\chi, \rho(x)_{[x/\omega]}) \geq \lambda$ that restricts the set of relevant direct inference statements to just those that are rationally representative to at least degree λ. The function f is a considerable simplification of matters, for what will appear in the premises are statements concerning drawn samples and the *known* statistical classes to which they belong together with knowledge that would rule-out particular statistical statements from use (Kyburg and Teng, 1999; Wheeler, 2004; Wheeler and Damásio, 2004).

Once this is in place, then ψ is understood to express the pair $\langle \chi, [l, u] \rangle$, which is the assignment of evidential probability $[l, u]$ to χ given $\varphi_1, \ldots, \varphi_n$, and this probability is the proportion of the models in $\mathcal{M}$ of $\varphi_1^1, \ldots, \varphi_n^1$ that are also models of χ. For details, see §4 and (Kyburg and Teng, 2001).

This approach does not resolve the controversy surrounding the fiducial argument, of course, but it does offer a view about what is behind the controversy and also offers guidance over how to proceed in evaluating such a line of reasoning. On this view the problem is that we typically aren't given information sufficient to determine f, but rather may have partial information that we may use to rule out particular relevant statistical statements that would otherwise be applicable to χ. The novelty of the fiducial argument is to treat indirect inference as a form of direct inference, and we see that EP uses knowledge about a sample and its method to eliminate those reference statistics from which we have reason to think are not representative—that fail to be rationally representative to some specified degree, λ. The strategy for eliminating candidates is similar to the principle of Richness, Specificity, and Strength in §4, and is a guiding principle of EP: exploit the knowledge you have to eliminate dubious or too-general statistics, then accept the uncertainty that remains.

Second-Order EP and the Fiducial Argument.

To interpret Schema (1.1) as a second-order EP probability about a particular fiducial argument is to follow the general strategy of second-order EP in §4.3.3 but with one exception. To effect the (partial) restriction of relevant statistical statements that is ideally the task performed by the rational representativeness function f, we need to exploit information in our knowledge base to eliminate candidate statistics and the truth values of these statements should be held constant rather than allowed to vary according to the theory of counter-factual EP presented in §4.3.2. There will be meta-linguistic machinery for setting thresholds and eliminating sentences, similar to the machinery for applying the EP rules of Richness, Specificity, and Strength. But, supplied with such machinery, we can propose a semantics for robust fiducial arguments simply in terms of a conditional second-order EP probability, where the conditioning event is all information in the knowledge-base that determines the relevant statistics whose reference classes are also rationally representative of the statement χ of the first-order EP inference.

Chapter 6
Bayesian Statistical Inference

Bayesian statistics is much more easily connected to the inferential problem of Schema (1.1) than classical statistics. The feature that distinguishes Bayesian statistical inference from classical statistics is that it also employs probability assignments over statistical hypotheses. It is therefore possible to present a Bayesian statistical procedure as an inference concerning probability assignments over hypotheses. Recall that we called the inference of probability assignments over data on the assumption of a statistical hypothesis direct. Because in Bayesian inference we derive a probability assignment over hypotheses on the basis of data, it is sometimes called indirect inference.

The basic structure of Bayesian statistical inference extends to Bayesian inference outside the statistical domain, as it is for example used in philosophical and psychological modelling. Any such inference starts with a combination of probability assignments, from which further probability assignments are derived using Bayes' theorem. And because this is a theorem of probability, Bayesian inference is also very close to the inferences dealt with in §2, which are based solely on the axioms as well. However, links to the standard semantics and to the role of Bayesian inference in philosophy and psychology are not discussed here.

6.1 Background

While, as illustrated in Howson and Urbach (1993), Bayesian inference has made its mark on numerous domains of probabilistic inference in science, its earliest application, in the work of Bayes himself, is in statistical inference (Earman, 1992). Nevertheless, in the development of statistical science during the previous century, Bayesian inference played a minor role. It is only with the development of the proper computational tools and machinery that this has taken a turn for the better. Presently, Bayesian statistics enjoys a growing popularity, as witnessed by numerous good introductory textbooks Bernardo and Smith (2000); Press (2003); Gelman et al. (2003)

R. Haenni et al., *Probabilistic Logics and Probabilistic Networks*, Synthese Library 350,
DOI 10.1007/978-94-007-0008-6_6, © Springer Science+Business Media B.V. 2011

It should be emphasized that not everyone is equally enthusiastic about the recent popularity of Bayesian techniques. Mayo (1996) argues forcefully against the Bayesian use of priors and the Bayesian reliance on likelihoods as the sole mediator between evidence and hypotheses, pointing to counterintuitive consequences of the latter view, e.g., on the issue of optional stopping. And even within the Bayesian camp itself, views diverge widely on what exactly characterizes Bayesian statistics: some authors side with strict subjectivists like De Finetti, interpreting all probabilities as degrees of belief, others propose to interpret at least some probabilities as objective, or pertaining to frequencies, and adhere to so-called mixed Bayesianism Jeffrey (1992). But none of these matters need concern us here. All Bayesians agree on the use of Bayes' theorem as a means of reasoning statistically, and the task for us here is to capture this in Schema (1.1).

Let $\Omega_H \times \Omega_D$ be the combination of a partition of hypotheses and a sample space, and let P be a probability assignment over this space. We can then define Bayesian statistical inference as follows. See Barnett (1999) and Press (2003) for more detail.

Definition 14 (Bayesian Statistical Inference). Assume $P(H_j)$, the prior probabilities assigned to a finite number of hypotheses H_j with $0 < j \leq n$, and $P(D|H_j)$, the probability assigned to the data D conditional on the hypotheses, called the likelihoods. Bayes' theorem determines that

$$P(H_j|D) = P(H_j) \frac{P(D|H_j)}{P(D)}. \tag{6.1}$$

Bayesian statistical inferences is the transition from the prior $P(H_j)$ to the posterior $P(H_j|D)$.

Credence intervals and Bayesian estimations can all be derived from the posterior distribution over the statistical hypotheses in the model.

It may be noted that the probability of the data $P(D)$ can be hard to compute. One possibility is to use the law of total probability,

$$P(D) = \sum_j P(H_j)P(D|H_j).$$

But often the interest is only in comparing the ratio of the posteriors of two hypotheses. By Bayes' theorem we have

$$\frac{P(H_1|D)}{P(H_0|D)} = \frac{P(H_1)P(D|H_1)}{P(H_0)P(D|H_0)},$$

and if we assume equal priors $P(H_0) = P(H_1)$, we can use the ratio of the likelihoods of the hypotheses, the so-called Bayes factor, to compare the hypotheses.

Two further remarks are in order. First note that the terms appearing in the above equations, both hypotheses H_j and data D, refer to sets in the space $\Omega = \Omega_D \times \Omega_H$. So we associate each statistical hypothesis H_j with an entire sample space Ω_D, and similarly every assignment d is associated with subsets $d \times \Omega_H$. This reflects that every statistical hypothesis is logically consistent with every sample d. However, within the sample space Ω_D associated with the hypothesis H_j, there is often a designated set in Ω_D with probability 1. This set is determined by the law of large numbers, applied to the probability assignment or likelihood function, $P(D|H_j)$, as prescribed by the hypothesis. As explicated by Gaifman and Snir (1982), the infinitely long sample sequences in this set are often called random.[1]

Second, we want to distinguish between the above applications of Bayes' theorem and applications of what is often called Bayes' rule:

$$P_d(H_j) = P(H_j|d). \tag{6.2}$$

The rule is an epistemological principle, relating two different probability functions that pertain to different epistemic states of an agent, at different points in time. Bayes' theorem, by contrast, is a mathematical fact. It puts a constraint on a probability assignment over an algebra, much like the proof theory of a logic sets constraints on truth valuations over an algebra. In the present section we will concentrate fully on the latter, because we are not considering an explicitly epistemic application of Bayesian inference.

6.2 Representation

The derivation of the posterior $P(H_j|d)$ from the prior $P(H_j)$ and the likelihoods $P(d|H_j)$ can be represented straightforwardly in Schema (1.1). Abbreviating the likelihoods as $P(d|H_j) = \theta_j(d)$, we can write:

$$\forall j \leq n: H_j^{P(H_j)}, (d|H_j)^{\theta_j(d)} \models (H_j|d)^{P(H_j|d)}. \tag{6.3}$$

In words, the schema combines probabilistic premises, namely the priors and likelihoods of hypotheses, to arrive at probabilistic conclusions, namely a conditional posterior over the hypotheses. Note that all arguments in Schema (6.3) are at bottom restrictions to a class of probability assignments, or models for short. Hence the inference of Schema (6.3) follows quite naturally from the standard semantics presented in §2: the probabilistic conclusions are drawn solely on the basis of the axioms of probability, which serve as derivation rules.

---

[1] In a sense this particular set determines the observational content of the statistical hypotheses. In (Romeijn, 2005) it is argued that we can also identify H_j with that particular set in Ω_D that is given probability 1 by the probability assignment associated with H_j.

6.2.1 Infinitely Many Hypotheses

The representation of Bayesian statistical inference in Schema (1.1) is not entirely unproblematic. An important restriction is the number of statistical hypotheses that can be considered in Schema (6.3).

Grouping Hypotheses.

Many statistical applications do not employ a finite number of hypotheses H_j, but a continuum of hypotheses H_θ. Now the mathematics of this is quite unproblematic: we collect the infinity of hypotheses in a space with a metric, $\theta \in \Theta$, over which we define a probability distribution $P(\theta)d\theta$ and a likelihood function $P(d|H_\theta) = \theta(d)$. In Schema (6.3) we can write this down as

$$\forall \theta : H_\theta^{P(\theta)d\theta}, (d|H_\theta)^{\theta(d)} \models (H_\theta|d)^{P(\theta|d)d\theta}. \tag{6.4}$$

However, when cast in terms of the schema like that, the inference involves an uncountable infinity of premises and conclusions, which means that we must assume an unusually rich logical language.

One reaction to this is to bite the bullet and define a logical language of that size.[2] But if framing the statistical inferences in a logical schema necessitates the use of such heavy material, we may ask whether the frame is not a bit too heavy for the picture we are trying to present. Both for conceptual and computational reasons, we will not develop the idea of directly employing a continuum of hypotheses with a density-valued probability assignment over it. Instead we may again employ a finite number n of statistical hypotheses H_j, each of which is composed of a set of hypotheses H_θ, for example according to

$$H_j = \left\{ H_\theta : \theta \in \left[\frac{j-1}{n}, \frac{j}{n} \right) \right\},$$

with the interval closed on both sides for $j = n$. Choosing $\theta_j = \frac{2j-1}{2n}$ we can then approximate the continuous model arbitrarily close by increasing n.

De Finetti's Representation.

Another possible reaction to the uncountable infinity involved in using statistical hypotheses is to translate the schema with statistical hypotheses into something more manageable. In particular, we will have to restrict the statistical inferences

---

[2] If we define statistical hypotheses in terms of sets of infinite sequences of observations, as suggested in Footnote 1, then it seems that we can employ a somewhat smaller logical language, which only involves a countable infinity of premises. But it leads us too far from the main line of this section to make this precise.

to a specific class. First, we say that the data consist of assignments to binary propositional variables D_i, denoted $d^e = d_1^{e(1)} \cdots d_m^{e(m)}$. Further, we restrict attention to statistical hypotheses that fix binomial distributions for the data, meaning that separate assignments $d_i^{e(i)}$ are associated with independent fixed chances: $P\left(d_{m+1}^1 \mid d^e, h_\theta\right) = \theta \in [0,1]$.

Finally, for convenience, we focus on specific predictions $P(d^{e'} \mid d^e)$, in which e' is a vector of length $m' > m$. These predictions can be derived from the posterior probability assignments $P(H_\theta \mid d^e)$ and the likelihoods $\theta(d^{e'})$ for $d^{e'}$ by the law of total probability. These predictions fit the following schema:

$$\forall \theta : H_\theta^{P(H_\theta)}, (d^e \mid H_\theta)^{\theta(d^e)}, (d^{e'} \mid H_\theta)^{\theta(d^{e'})} \models (d^{e'} \mid d^e)^{P(d^{e'} \mid d^e)}, \tag{6.5}$$

where the likelihoods of H_θ for d^e is the product of the likelihoods of the separate assignments $d_i^{e(i)}$. With $m_1 = \sum_i e(i)$ we have $\theta(e) = \theta^{m_1}(1 - \theta)^{m-m_1}$. Note, however, that this schema has infinitely many premises.

As de Finetti (1937) shows, we can represent any statistical inference of (6.5) in terms of probability assignments over finitely many samples. The premises in (6.5) can be replaced by another set of premises, which do not mention the hypotheses H_θ, but instead refer to the so-called exchangeability of the probability P for samples d^e and $d^{e'}$. An exchangeable probability of the sample d^e is invariant under permutations of the assignments e:

$$\pi(\langle e(1), \ldots e(i), \ldots, e(m)\rangle) = \langle e(i), \ldots e(1), \ldots, e(m)\rangle$$

for any i. The new inference then becomes

$$\forall \pi : \left(d^{\pi(e')}\right)^{P(d^{e'})} \models \left(d^{e'} \mid d^e\right)^{P(d^{e'} \mid d^e)}. \tag{6.6}$$

where $\pi(e')$ refers to any order permutation of the elements in the vector e'. The salient point is that we can infer the very same predictions $P(d^{e'} \mid d^e)$ that were determined in (6.5) from a probability assignment for which the samples $d^{e'}$ and any order permutation of the data $d^{\pi(e')}$ are equally probable.

This entails the restriction on the set of probability assignments induced by the set of premises $\{H_\theta^{P(\theta)}, (d^e \cap H_\theta)^{\theta(d^e)}, (d^{e'} \cap H_\theta)^{\theta(d^{e'})}\}$ is exactly the same as the restriction induced by the premises in (6.6). But this latter restriction is set in strictly finite terms: since the data is finite, there are only finitely many permutations of the data elements. The representation theorem of de Finetti thus allows us to dispose of the infinity of premises that is involved in the standard Bayesian statistical inference.

Carnapian Inductive Logic.

Carnap (1952, 1980), Kuipers (1978), and more recently Paris and co-workers ((Paris, 1994; Nix and Paris, 2007) and further references therein) consider special cases of Schema (6.6) in what has become known as inductive logic. The general idea of inductive logic is that invariances under the permutations π of Schema (6.6) can be motivated by independent rationality criteria or principles of logic. The probabilistic conclusions, generally predictions $P(d^{e'}|d^e)$, thus follow analytically.

A well-known inductive logic is given by the so-called continuum of inductive methods. Assuming invariance under order permutation π as explicated in the foregoing, the invariance under permutations of the values,

$$\pi(\langle e(1),\ldots e(i),\ldots,e(m)\rangle) = \langle 1 - e(1),\ldots 1 - e(i),\ldots,1 - e(m)\rangle,$$

and the principle of restricted relevance, which comes down to the assumption that $P\left(d^1_{m+1}|d^e\right)$ is a linear function of m_1 only, we can derive the following predictions,

$$P\left(d^1_{m+1}|d^e\right) = \frac{m_1 + \frac{\lambda}{2}}{m + \lambda}. \tag{6.7}$$

Here m_1 is as before and λ is a free parameter. With de Finetti's representation, we can also capture these predictions in Schema (6.5). The value of λ is then determined by the distribution $P(H_\theta)d\theta$.[3]

6.2.2 Interval-Valued Priors and Posteriors

Until now we have focused on sets of probability assignments that can be defined by restrictions in terms of sharp probability assignments. But we may also represent restrictions in terms of intervals of probability assignments. We can use this to represent a wider class of Bayesian statistical inferences in Schema (1.1).

Walley (Lindley, 1963) discusses interval restrictions in the context of the above multinomial hypotheses and their associated Carnapian predictions. As said, a sharp distribution over multinomial hypotheses leads to an exchangeable prediction rule. Walley points out that we can also allow for interval-valued assignments to statistical hypotheses, and that such valuations can be dealt with adequately by considering a class of prior distributions over the statistical hypotheses instead of a single and sharp-valued prior distribution. These interval-valued assignments are useful in studying the sensitivity of statistical results to the prior probability that is chosen.

Furthermore, classes of prior distributions are associated with ranges of prediction rules. As Skyrms (1993) and Festa (2006) show, we can use such ranges of rules to constitute so-called hyper-Carnapian prediction rules: each of the the prediction rules reacts to the incoming sample separately, but each prediction rule is also assigned a second-order probability, and this latter probability assignment is updated according to the data as well. The resulting prediction rules have quite interesting properties. Most notably, they can be used for deriving predictions that incorporate

---

[3] In particular, depending on the value of λ we must adopt a specific symmetric Dirichlet distribution. See Festa (1993).

analogical predictions. However, it is not easy, nor is it in our opinion very insightful, to represent the higher-order probability assignments over Carnapian rules in terms of the scheme of Schema (1.1). In this book we will therefore not study the hyper-Carnapian prediction rules.

6.3 Interpretation

We can interpret Schema (1.1) straightforwardly as a Bayesian statistical inference, as long as the inference concerns probability assignments over a sample space and a model, $\Omega_D \times \Omega_H$. One type of premise concerns the data and the statistical model: we restrict ourselves to those probability assignments for which the likelihoods of the hypotheses on the data, $P(D|H_j)$, have specific values. The other premise is the prior probability assignment: we restrict the set of probability assignments to those for which the marginal probability over hypotheses $P(H_j)$ is equal to specific values. From these two restrictions and the assignment d to the data variable D we can derive, according to the progicnet programme, a further restriction on the posterior probability $P(H_j|d)$.

6.3.1 Interpretation of Probabilities

Bayesian statistics is very often associated with the so-called subjective interpretation of probability. Thus far we have not determined the interpretation of the probability assignments featuring in the inference. We want to briefly comment on the issue of interpreting probability here.

First, as probabilistic logicians we are not necessarily tied to a specific interpretation of the probability functions, just like classical deductive logicians are not necessarily tied to Tarski's or some other truth definition as an interpretation of the truth values in their inferential schemes. This is fortunate, because the debate on interpretations of probability is very complicated, and it easily muddles the discussion on probabilistic inference, especially Bayesian inference. On the whole, our present concerns are not with these particular interpretative issues, but with the logic and its formal semantics.

If we are interpreting probabilities, we should distinguish between the probability assigned to statistical hypotheses H_j and to samples or data d. With regard to the former, note that the schema concerns statistical hypotheses in two ways: on the one hand the hypotheses are expressions in the language, on the other hand they determine a probability assignment over the language. Now in a standard Bayesian account, probability assignments over hypotheses seem most naturally understood as epistemic, meaning that they pertain to degrees of belief. Note that this does not yet mean that probabilities of hypotheses are subjective, because we may also try to determine the epistemic probability assignment by means of objective criteria.

As for the probability assigned to samples, the likelihoods, we may give it either a physical interpretation, meaning that the probabilities refer to aspects of the physical world, or an epistemic interpretation. Arguably, the probability assignments as prescribed by the hypotheses are best interpreted as a physical notion, because hypotheses are statements about the world. At the same time the likelihoods of the hypotheses seem best interpreted as epistemic, because they fulfill an evidential role.

6.3.2 Bayesian Confidence Intervals

There is another, rather natural interpretation of interval probabilities in the context of Bayesian statistical inference. We discuss it briefly here to clarify its relation to Schema (1.1), and to explain why we will not pursue this interpretation in what follows.

Recall that the continuum of hypotheses H_θ concerns probability assignments over data D via the parameter θ, and that we have defined probability density functions over this parameter, $P(H_\theta|D)$. From these density functions we can define Bayesian confidence intervals, or credence intervals, for the values of θ. Each interval $\theta \in [l, u]$ is associated with a posterior probability, or credence,

$$P(h_{[l,u]}|D) = \int_l^u P(H_\theta|D)d\theta.$$

Now we may fix u and l such that

$$\int_0^l P(H_\theta|D)d\theta = \int_u^1 P(H_\theta|D)d\theta = 2.5\%.$$

In that case we have $P(h_{[l,u]}|D) = 0.95$, and we can then say that we are 95% certain that the real value of θ lies within the interval $[l, u]$. From the knowledge of a sharp-valued probability assignment over θ, we might even construct an interval-valued probability assignments for θ, written $\theta \in [l, u]$.

Credence intervals might lead us to reconsider the representation of the Bayesian statistical inferences in Scheme (6.3), and bring Bayesian statistical inference closer to classical statistical inference. Assuming a uniform prior probability density $P(H_\theta) = 1$, the corresponding interval-valued assignment is $\theta \in [.025, .975]$. But after incorporating a specific sample d, we obtain the posterior density function $P(H_\theta|d)$. It may so happen that the corresponding interval-valued assignment is $\theta \in [.08, .13]$, meaning that conditional on obtaining the data d, the 95% credence interval shrinks. The question arises whether we can somehow interpret the interval-valued probabilities in Schema (1.1) as such credence intervals.

Unfortunately, we cannot. The inferences of interval-valued probabilities in Schema (1.1), when interpreted as credence intervals, are elliptic, that is, they omit certain premises. More precisely, the fact that $\theta \in [.025, .975]$ does not fix the detailed shape of the prior probability density $P(H_\theta)$. But we need this detailed shape to arrive at the specific conditional credence interval $\theta \in [.08, .13]$. It is not possible to rely just on the rules of inference of Schema (1.1) for these credence intervals. An interpretation of Schema (1.1) in terms of the credence intervals deriving from Bayesian statistical inference is therefore not possible.

Chapter 7
Objective Bayesian Epistemology

We now turn to the final semantics for probabilistic logic that we consider in this book: the objective Bayesian semantics. In §7.1 we provide an introduction to objective Bayesian epistemology, which holds that the strengths of an agent's beliefs should be representable by certain probability functions. The key task is to say which probability functions are appropriate (§7.1.1). Objective Bayesian epistemology holds that an agent's evidence should constrain her degrees of belief (§7.1.2), and that where evidence underdetermines appropriate degrees of belief they should be as equivocal as possible. §7.1.3 and §7.1.4 explain the equivocation requirement on propositional and predicate languages respectively.

In §7.2 we see that questions for objective Bayesianism can be formulated as questions of the form of Schema (1.1). On the other hand, Schema (1.1) can be interpreted by appealing to objective Bayesian epistemology (§7.3). Hence objective Bayesianism plugs into the general framework of the book.

7.1 Background

According to objective Bayesian epistemology, an agent's rational degrees of belief are determined largely by the extent and limitations of the propositions $\mathcal{E}$ that she takes for granted (Williamson, 2009b). $\mathcal{E}$ is called the agent's *total evidence*, or *epistemic background*, and includes her background knowledge, observations, theoretical assumptions etc.—everything that is not under question in her current operating context.

The agent's evidence determines the strengths of her beliefs in two ways. First, the agent's degrees of belief should be compatible with this evidence: e.g., if she grants φ, she should fully believe it; if her evidence consists just of a narrowest reference-class frequency pertinent to φ then she should believe it to the extent of that frequency; if her granted physical theory determines the chance of φ then she should believe φ to the extent of that chance; if her granted physical theory says that there is a symmetry between φ and φ' which results in them having the same

R. Haenni et al., *Probabilistic Logics and Probabilistic Networks*, Synthese Library 350, DOI 10.1007/978-94-007-0008-6_7, © Springer Science+Business Media B.V. 2011

chance then she should award them the same degree of belief.[1] Second, her degrees of belief should otherwise equivocate between the basic possibilities that she can countenance—she should not adopt extreme degrees of belief unless forced to by her evidence.

7.1.1 Determining Objective Bayesian Degrees of Belief

Importantly, objective Bayesianism—in common with other versions of Bayesian epistemology—holds that the agent's rational degrees of belief are representable by a probability function; thus her degree of belief in φ can be measured by a single number in the unit interval. This position can be motivated by betting considerations (Ramsey, 1926; de Finetti, 1937) or by derivation from intuitive principles (Cox, 1946). Many Bayesians also agree with the objective Bayesian tenet that the agent's probability function should satisfy constraints imposed by $\mathscr{E}$.[2] The motivation behind this tenet is that an agent's degrees of belief will be used as a basis for inference and decision; success in these tasks require calibration with physical probability.

What separates objective Bayesianism from other varieties of Bayesianism is its further insistence on equivocal degrees of belief: an agent's probability function should be a probability function, from those satisfying constraints imposed by evidence, that is sufficiently non-extreme. This is because extreme degrees of belief tend to trigger risky actions while equivocal degrees of belief are associated with less risk, and it is prudent only to take on risk to the minimum extent warranted by available evidence (Williamson, 2007b).

The general recipe for objective Bayesian assignment of belief proceeds as follows. The evidence $\mathscr{E}$ determines a set $\mathbb{E}$ of probability functions which are compatible with $\mathscr{E}$. The agent should then have degrees of belief that are representable by a probability function $P_{\mathscr{E}}$ which is in $\mathbb{E}$ and is sufficiently close to an *equivocator*, a probability function $P_=$ that is maximally equivocal over the agent's language. What counts as *sufficiently* close depends in part on pragmatic factors such as the numerical accuracy required for applications; since a question of the form Schema (1.1) gives little clue as to pragmatic factors and applications it is natural in the context of probabilistic logic to identify *sufficiently close* and *closest* (Williamson, 2010, Chapter 7), and we shall adopt this simplification here: $P_{\mathscr{E}}$ should be a function in $\mathbb{E}$ that is closest to the equivocator $P_=$. To flesh out this procedure, we need to say a bit

---

[1] For an agent's degrees of belief to be rational, it is sufficient that they be rationally determined from her evidence $\mathscr{E}$, that this evidence consists of propositions that the agent takes for granted, and that the agent is rational to take these propositions for granted given her purposes at the time. The truth of the propositions in $\mathscr{E}$ is not required. Hence the propositions in $\mathscr{E}$ need not count as *knowledge* in the philosophical sense that presupposes truth.

[2] Some Bayesians restrict this to the case in which the evidence in $\mathscr{E}$ takes the form of sentences in the domain of the probability function (in which case the constraints are usually just that these sentences should be fully believed). The objective Bayesian does not impose this restriction.

more about how $\mathbb{E}$ is determined by $\mathscr{E}$, and more precisely characterise the notions of *equivocator* and *closeness* to the equivocator.

7.1.2 Constraints on Degrees of Belief

First we shall say a few words about how $\mathbb{E}$ is determined by $\mathscr{E}$. The idea here is that $\mathscr{E}$ imposes constraints χ that an agent's degrees of belief should satisfy. This set of constraints can be used to characterise $\mathbb{E}$, the set of probability functions that are compatible with $\mathscr{E}$.

Definition 15 (Directly Transferred Constraints). The set of *directly transferred constraints* is the smallest set χ such that:

- if $\mathscr{E}$ implies that $chance(\varphi) \in X$ then $P(\varphi) \in X$ is in χ (in particular, if φ is in $\mathscr{E}$ then $P(\varphi) = 1$ is in χ); more generally, substitute P for $chance$ when transferring constraints to χ,
- if $\mathscr{E}$ implies that $freq_U(V) \simeq x$ (which says 'the frequency of V in reference class U is approximately x') and $Vt \leftrightarrow \varphi$ and Ut and there is no narrower U' for which this holds, then $P(\varphi) = x$ is in χ; more generally, let known narrowest-reference-class frequencies transfer to constraints on P in χ.[3]

Let $\mathbb{P}$ be the set of all probability functions and let $\mathbb{P}_\chi$ be the set of probability functions that satisfy χ, the directly transferred constraints. There are two reasons why we can't just set $\mathbb{E} = \mathbb{P}_\chi$. First, it is sometimes unreasonable to require that degrees of belief should satisfy the same constraints as physical probabilities (frequencies, chances). For example, suppose the agent knows just that the truth of φ has already been decided (φ asserts the occurrence of an event in the past, say). This implies that $(chance(\varphi) = 0) \vee (chance(\varphi) = 1)$. But this does not impose the constraint $(P_\mathscr{E}(\varphi) = 0) \vee (P_\mathscr{E}(\varphi) = 1)$. Arguably, it does not impose any constraint at all, in which case the objective Bayesian method will yield $P_\mathscr{E}(\varphi) = 1/2$. In general, objective Bayesianism's epistemological basis and its advocacy of non-extreme probabilities motivate taking the *convex hull* of directly transferred constraints, denoted by $\langle \mathbb{P}_\chi \rangle$ (Levi, 1980, §9; Williamson, 2005a, §5.3).[4]

The second reason why we can't just set $\mathbb{E} = \mathbb{P}_\chi$ is that $\mathbb{P}_\chi$ might be empty, i.e., the constraints in χ might be inconsistent (perhaps because there is more than one narrowest reference class frequency pertinent to φ). Of course an agent is entitled

---

[3] Depending on the closeness of the approximation $freq_U(V) \simeq x$ (e.g., depending on the sample size) one might relax this constraint to $P(\varphi) = [x - \delta, x + \delta]$ for some suitable δ. One can use the principles of evidential probability to explicate the way in which narrowest-reference-class frequencies constrain P; see Wheeler and Williamson (2010) for more details.

[4] Note that objective Bayesianism admits *structural* as well as directly transferred constraints: structural constraints are imposed by qualitative evidence, e.g., evidence of causal, ontological, logical, or hierarchical relationships. As explained in Williamson (2005a), structural constraints take the form of equality constraints. Since structural constraints are not central in the context of probabilistic logic, we shall focus solely on directly transferred constraints in this book.

to hold beliefs in such a circumstance, so $\mathbb{E}$ cannot be empty. Hence some kind of consistency maintenance procedure needs to be invoked. One such procedure involves requiring that $P_{\mathscr{E}}$ be compatible with some maximal consistent subset of χ. Let $\mathbb{P}_{\chi}^{\uplus} = \bigcup_{\chi^+} \mathbb{P}_{\chi^+}$ where χ^+ ranges over the maximal consistent subsets of χ. Then,

Definition 16 (Compatible). The set $\mathbb{E}$ of probability functions that are *compatible* with $\mathscr{E}$ is defined by $\mathbb{E} \overset{\mathrm{df}}{=} \left\langle \mathbb{P}_{\chi}^{\uplus} \right\rangle$.

Thus the probability functions compatible with the agent's evidence $\mathscr{E}$ are those that are in the convex hull of the set of probability functions that satisfy constraints directly transferred from $\mathscr{E}$, if these constraints are consistent, or in the convex hull of maximal consistent subsets of the directly transferred constraints if not.

Next we turn to the notions of equivocator and closeness to the equivocator. These need to be defined relative to a particular language. We will consider two kinds of language here: a propositional language and a predicate language.

7.1.3 Propositional Languages

The simplest case is that in which the agent's language is a finite propositional language built on assignments $a_1, \ldots, a_n$ to propositional variables. The natural notion of distance between two probability functions on this domain is the cross entropy divergence $d(P,Q) = \sum_{\alpha} P(\alpha) \log P(\alpha)/Q(\alpha)$, where the α range over the atomic states. Note that $0 \log 0$ is taken to be 0 and that d is not a distance function in the usual mathematical sense because it is not symmetric and does not satisfy the triangle inequality. The natural equivocator is defined by $P_=(\alpha) = 1/2^n$; it equivocates between atomic states $a_1^{e_1} \wedge \cdots \wedge a_n^{e_n}$ (where $e_1, \ldots, e_n \in \{0,1\}$). Let $\downarrow\mathbb{E}$ be the subset of $\mathbb{E}$ that contains those probability functions closest to the equivocator, $\downarrow\mathbb{E} \overset{\mathrm{df}}{=} \{P \in \mathbb{E} : d(P,P_=) \text{ is minimised}\}$. (If there is no such function then one can take $\downarrow\mathbb{E}$ to be the set of functions that are *sufficiently* close to the equivocator. As mentioned above, as to which functions are sufficiently close will depend on contextual considerations such as measurement accuracy. This case is not central to our concerns here though: since $\mathbb{E}$ is convex and d is a strictly convex function, if $\mathbb{E}$ is closed—as it is in the context of a propositional language and a probabilistic logic that admits closed-interval constraints on sentences—then $\downarrow\mathbb{E}$ is a singleton.) The objective Bayesian recipe is then to choose $P_{\mathscr{E}} \in \downarrow\mathbb{E}$. Now minimising distance with respect to the equivocator is equivalent to maximising *entropy* $H(P) = -\sum_{\alpha} P(\alpha) \log P(\alpha)$. Hence on this finite domain we have the *Maximum Entropy Principle*:

Maximum Entropy Principle: On a finite propositional language the agent's degrees of belief should be representable by a probability function $P_{\mathscr{E}}$, from all those compatible with $\mathscr{E}$, that has maximum entropy (Jaynes, 1957).

7.1.4 Predicate Languages

If the language is a predicate language then the picture is similar, but slightly more complicated.[5] We shall assume that the language has finitely many predicate symbols, countably many constant symbols t_i, but no function symbols or equality. Let $\mathcal{L}_n$ be the finite sublanguage involving all the predicate symbols but only the constant symbols $t_1, \ldots, t_n$. Create an ordering $a_1, a_2, \ldots$ of the atomic sentences— sentences of the form Ut where U is a predicate or relation symbol and t is a tuple of constants of corresponding arity—ensuring that those atomic sentences of $\mathcal{L}_m$ occur in the ordering before those that are expressible in $\mathcal{L}_n$ but not in $\mathcal{L}_m$, for $m < n$. We then consider the n-divergence on $\mathcal{L}_n$,

$$d_n(P,Q) \stackrel{\mathrm{df}}{=} \sum_{e_1,\ldots,e_n=0}^{1} P\left(a_1^{e_1} \wedge \cdots \wedge a_n^{e_n}\right) \log \frac{P\left(a_1^{e_1} \wedge \cdots \wedge a_n^{e_n}\right)}{Q\left(a_1^{e_1} \wedge \cdots \wedge a_n^{e_n}\right)}.$$

Here, as before, a_i^1 signifies a_i and a_i^0 signifies $\neg a_i$. As before we take $0\log 0$ to be 0. We will not need to consider the case $x\log x/0$ and will simply assume in what follows that Q is never zero.

On a predicate language the equivocator is defined by

$$P_=\left(a_1^{e_1} \wedge \cdots \wedge a_n^{e_n}\right) \stackrel{\mathrm{df}}{=} 1/2^n,$$

for all n. Thus

$$d_n(P,P_=) = \sum_{e_1,\ldots,e_n=0}^{1} P\left(a_1^{e_1} \wedge \cdots \wedge a_n^{e_n}\right) \log \left[2^n P\left(a_1^{e_1} \wedge \cdots \wedge a_n^{e_n}\right)\right].$$

Letting $d(P,Q) = \lim_{n\to\infty} d_n(P,Q) \in [0,\infty]$, one might try, as before, to choose $P_{\mathcal{E}} \in \{P \in \mathbb{E} : d(P,P_=)$ is minimised$\}$. But as it stands this does not adequately explicate the concept of closeness to the equivocator, because in the case of a predicate language there are probability functions P, Q such that although one is intuitively closer to the equivocator than the other, $d(P,P_=) = d(Q,P_=)$. Suppose for example that $\mathcal{E}$ imposes the constraints $P\left(a_n|a_1^{e_1} \wedge \cdots \wedge a_{n-1}^{e_{n-1}}\right) = 1$ for all $n \geq 2$. Thus only $P(a_1)$ is unconstrained. Then $\mathbb{P}_\chi$ is non-empty, closed and convex, so $\mathbb{E} = \mathbb{P}_\chi$, but $d(P,P_=) = \infty$ for all $P \in \mathbb{E}$. Yet intuitively there is a unique function in $\mathbb{E}$ that is closest to the equivocator, namely the function that sets $P(a_1) = 1/2$ and $P\left(a_n|a_1^{e_1} \wedge \cdots \wedge a_{n-1}^{e_{n-1}}\right) = 1$ for all $n \geq 2$. Accordingly we introduce the following definition.

---

[5] Note that the approach advocated here is very different to that of Williamson (2007a) and Williamson (2008b). The merit of this new approach is that it permits a uniform treatment of propositional and predicate languages. We assume, as is normal with discussions of probabilities over predicate languages, that each element of the domain is picked out by some constant symbol; see Paris (1994) on this point.

Definition 17 (Closer). For probability functions P, Q, R defined on a predicate language, we say P is *closer* than Q to R iff there is some number N such that for all $n \geq N$, $d_n(P,R) < d_n(Q,R)$.

Note that in the above example, the function that is intuitively closer to the equivocator is indeed deemed to be closer (N can be taken to be 1 in this case). Here is another example:

Example 10. Suppose $\mathbb{E} = \{P \in \mathbb{P} : P(\forall x U x) = c\}$ for some fixed $c \in [0,1]$. (It is usual to understand a constraint of the form $P(\forall x U x) = c$ as $\lim_{n \to \infty} P(U t_1 \wedge \cdots \wedge U t_n) = c$).[6] We have that $d(P, P_=) = \infty$ for all $P \in \mathbb{E}$. Define P by

$$P(U t_1) = \frac{c+1}{2}$$

$$P\left(U t_{i+1} | U t_1^{e_1} \wedge \cdots \wedge U t_i^{e_i}\right) = \begin{cases} \frac{(2^{i+1}-1)c+1}{(2^{i+1}-2)c+2} & : \quad e_1 = \cdots = e_i = 1 \\ \frac{1}{2} & : \quad \text{otherwise} \end{cases}$$

Then P is the member of $\mathbb{E}$ that is closest to the equivocator.

The objective Bayesian protocol is then to seek a probability function that is closest to the equivocator in the sense of Definition 17. Let $\downarrow\mathbb{E}$ be those members of $\mathbb{E}$ that are closest to the equivocator $P_=$ in the sense of Definition 17. (Again, if there are no closest members we must consider those that are, from a pragmatic point of view, sufficiently close.) Then objective Bayesianism advocates choosing $P_{\mathscr{E}} \in \downarrow\mathbb{E}$, as in the case of a propositional language. (While in the case of propositional probabilistic logics considered here $P_{\mathscr{E}}$ is uniquely determined, this is not necessarily so with a predicate language. See Williamson (2008a) or Williamson (2010, Chapter 7) for discussion of uniqueness and other properties of objective Bayesianism on predicate languages.)

The following objection to the above choice of equivocator crops up from time to time in the literature (see, e.g., Dias and Shimony, 1981, §4). Suppose the agent has no evidence $\mathscr{E}$; then she will set her degrees of belief according to the equivocator. In particular $P_=(Br_{101} | Br_1 \wedge \cdots \wedge Br_{100}) = P_=(Br_{101}) = 1/2$. Thus if she observes a hundred ravens and finds them all to be black then her degree of belief that raven 101 will be black is the same as if she had observed no ravens at all. Learning from experience becomes impossible with this equivocator.

This objection fails because $P_=(Br_{101} | Br_1 \wedge \cdots \wedge Br_{100})$ does *not* represent the degree to which the agent would believe that raven 101 is black were she to observed a hundred ravens all black. Suppose the agent's initial evidence $\mathscr{E} = \emptyset$; then $\mathbb{E} = \mathbb{P}$ and indeed $P_{\mathscr{E}} = P_=$. Suppose that agent were subsequently to observe a hundred ravens and find them all black. Her new evidence is $\mathscr{E}' = \{Br_1 \wedge \cdots \wedge Br_{100}\}$. Her degree of belief that raven 101 is black given this evidence is $P_{\mathscr{E}'}(Br_{101})$. This is obtained by determining the $P \in \mathbb{E}'$ that is closest to $P_=$. To determine $\mathbb{E}'$ we need

---

[6] See Paris (1994) for instance. Note that this construal requires our earlier assumption that each member of the domain is picked out by some constant symbol in the language.

to isolate the constraints imposed by $\mathcal{E}'$. Clearly $\mathcal{E}'$ imposes the constraint $P(Br_1 \wedge \cdots \wedge Br_{100}) = 1$, but the evidence also suggests that the relative frequency of ravens being black is near 1, so by Definition 15 also constrains $P(Br_i)$ to be near 1 for unobserved r_i. Now $P_{\mathcal{E}'}$ must satisfy these constraints so $P_{\mathcal{E}'}(Br_{101})$ will end up being near 1 (which seems reasonable given the agent's lack of other evidence). Thus the agent will learn from experience after all. More generally, the objective Bayesian update on new evidence e does not always agree with the corresponding conditional probability $P_{\mathcal{E}}(\cdot|e)$. One of the conditions for agreement is that e be *simple* with respect to previous evidence $\mathcal{E}$, i.e., that the learning of e should *only* impose the constraint $P(e) = 1$. In this case, e yields frequency information and so is not simple. See Williamson (2009a) for a fuller discussion of the differences between objective Bayesian updating and Bayesian conditionalisation.

7.1.5 Objective Bayesianism in Perspective

We see, then, that objective Bayesianism requires that the strengths of an agent's beliefs be representable by a probability function, from all those that are compatible with the propositions that she takes for granted, which is closest to a suitable equivocator on the domain. One can take much the same approach on both predicate and propositional languages.

Jakob Bernoulli may be have been the first to advocate all three of the central tenets of objective Bayesianism, namely that degrees of belief should be probabilistic, constrained by empirical evidence, and otherwise equivocal (Bernoulli, 1713, Part IV; Williamson, 2005a, §5.2). Jaynes (1957) provided the maximum-entropy foundations which work well on finite domains; how to handle uncountable domains remains a question of some controversy (Williamson, 2009b, §19). Uncountable domains are less central to logic and to the purposes of this book, and so will not be covered here. Nilsson (1986, §4) applied objective Bayesianism to probabilistic logic.

For readers wishing to learn more about objective Bayesian epistemology, Williamson (2009b) provides an introduction; more detailed expositions can be found in Rosenkrantz (1977); Jaynes (2003); Williamson (2010). It should be emphasised that objective Bayesian epistemology is the subject of some controversy: it has variously been accused, among other things, of being poorly motivated, of poorly handling qualitative evidence, of erroneously updating beliefs, of suffering from a failure to learn from experience, of being computationally intractable, of being susceptible to paradox, of yielding beliefs that are language-dependent, of being too objective, and of not being objective enough (Williamson, 2010, §1.3). See Williamson (2010) for a response to these criticisms.

Objective Bayesian epistemology as outlined here differs in an important sense from Bayesian statistical inference described in §6: here all probabilities are first-order, attaching to sentences of a logical language, while Bayesian statistical inference focuses on second-order probabilities, i.e., probabilities of probabilities.

Objective Bayesian statistical inference, which appeals to equivocal second-order probabilities, is an active research area in statistics. Objective Bayesianism differs from probabilistic argumentation (§3) inasmuch as it is concerned with the probability that a conclusion is true, rather than the probability that the conclusion's truth is forced by the truth of the premises. Evidential probability (§4) can be viewed as a formalism for determining $\mathbb{E}$, the set of probability functions that are compatible with the agent's evidence; objective Bayesianism goes beyond evidential probability in that it advocates equivocation as well as calibration with statistical evidence (Wheeler and Williamson, 2010).

7.2 Representation

The objective Bayesian faces the following kind of question: given evidence $\mathscr{E}$, to what extent should you believe a proposition ψ of interest? It is not hard to see how this might be phrased in terms of the Fundamental Question of Probabilistic Logic, Schema (1.1). Suppose for instance that $\mathscr{E} = \{\varphi_1, chance(\varphi_2) \in [0.6, 0.8], freq_U(V) \simeq 0.9, Vt \leftrightarrow \varphi_3, Ut\}$.[7] Then the objective Bayesian question can be phrased in the language of Schema (1.1) as follows:

$$\varphi_1^1, \varphi_2^{[0.6,0.8]}, \varphi_3^{0.9} \approx \psi^?$$

Thus objective Bayesian epistemology fits neatly into the framework of this book.

7.3 Interpretation

Conversely, a question of the form of Schema (1.1), $\varphi_1^{X_1}, \varphi_2^{X_2}, \ldots, \varphi_n^{X_n} \approx \psi^?$, can be given an objective Bayesian interpretation. Simply take the premises on the left-hand side to encapsulate constraints on P that are directly transferred from an agent's evidence (i.e., constraints in the set χ introduced above). Then the set Y to attach to ψ is the set $Y = \left\{ P_{\mathscr{E}}(\psi) : P_{\mathscr{E}} \in \downarrow \left\langle \mathbb{P}_\chi^\psi \right\rangle \right\}$. Assuming a finite domain, Y is a singleton.

Thus objective Bayesianism can be used to provide semantics for probabilistic logic. Let $\chi = \left\{ \varphi_1^{X_1}, \varphi_2^{X_2}, \ldots, \varphi_n^{X_n} \right\}$ and let v be ψ^Y. Then $\chi \approx v$ if and only if $P \in \downarrow \left\langle \mathbb{P}_\chi^\psi \right\rangle$ implies $P(\psi) \in Y$, i.e., if and only if $\downarrow \left\langle \mathbb{P}_\chi^\psi \right\rangle \subseteq \mathbb{P}_v$.

Example 11. Consider the question $\forall x U x^{3/5} \approx U t^?$. Objective Bayesian epistemology interprets the expression on the left-hand side as the constraint χ imposed on an

---

[7] To reiterate, while the propositions $\varphi_1, \varphi_2, \varphi_3$ are in the domain of the agent's probability function, expressions of the form $chance(\varphi_2) \in [0.6, 0.8], freq_U(V) \simeq 0.9, Vt \leftrightarrow \varphi_3, Ut$ are not assumed to be in this domain.

agent's degrees of belief by her evidence $\mathscr{E}$; this constraint is consistent and determines a closed convex set, so $\mathbb{E} = \left\langle \mathbb{P}_\chi^\uplus \right\rangle = \mathbb{P}_\chi = \{P \in \mathbb{P} : P(\forall x U x) = 3/5\}$. There is one function P in $\mathbb{E}$ that is closest to the equivocator, as described in Example 10. This function gives $P(Ut) = 4/5$ for each constant t. Hence the answer to the question is $4/5$.

Note that the monotonicity property of §1.3 fails, so entailment is not decomposable under the objective Bayesian semantics. Nevertheless, this notion of entailment still satisfies a number of interesting properties. In order to characterise these properties we shall need to appeal to some notions that are common in the literature on nonmonotonic logics—see, e.g., Kraus et al. (1990b); Makinson (2005, §3.2) and Hawthorne and Makinson (2007) for background. Construct an (uncountable) language $\mathscr{L}^\sharp$ by taking statements of the form φ^X as atomic propositions, where φ is a sentence of a propositional or predicate language $\mathscr{L}$ and X is a set of probabilities. $\mathscr{L}^\sharp$ can be thought of as the language of the Fundamental Question, Schema (1.1). A probability function P can then be construed as a valuation on $\mathscr{L}^\sharp$. Define a decomposable entailment relation $\models$ by $P \models \mu$ iff $P \in \mathbb{P}_\mu$ for sentence μ of $\mathscr{L}^\sharp$, where μ is composed from statements of the form φ^X by the usual logical operators. In particular, if μ is of the form φ^X, then $P \models \mu$ iff $P(\varphi) \in X$. For probability functions P and Q, define $P \prec Q$ iff P is closer to the equivocator than Q. Then $(\mathbb{P}, \prec, \models)$ is a *preferential model*: $\mathbb{P}$ is a set of valuations on $\mathscr{L}^\sharp$, $\prec$ is an irreflexive, transitive relation over $\mathbb{P}$, and $\models$ is a decomposable entailment relation. Moreover, this preferential model is *smooth*: if $P \models \mu$ then either P is minimal with respect to $\prec$ in $\mathbb{P}_\mu$ or there is a $Q \prec P$ in $\mathbb{P}_\mu$ that is minimal. Hence this model determines a *preferential consequence relation* $\mid\!\sim$ as follows: $\mu \mid\!\sim \nu$ iff P satisfies ν for every $P \in \mathbb{P}$ that is minimal among those probability functions that satisfy μ. We will be particularly interested in the case in which $\mathbb{P}_\mu = \mathbb{P}_\mu^\uplus = \left\langle \mathbb{P}_\mu^\uplus \right\rangle = \mathbb{E}$, the set of probability functions that are compatible with μ (which occurs for example where sentence μ of $\mathscr{L}^\sharp$ is consistent and quantifier-free and where all sets X occurring in μ are closed intervals). Call μ *regular* if it satisfies this property. Wherever μ and ν are regular sentences, $\mid\!\sim$ will agree with $\approx$, where the latter entailment relation is extended to $\mathscr{L}^\sharp$ in the obvious way. Consequently on regular sentences $\approx$ will satisfy the properties of preferential consequence relations, often called system-P properties—see, e.g., Kraus et al. (1990b):

Proposition 4 (Properties of Entailment). *Let $\models$ denote entailment in classical logic and let $\equiv$ denote classical logical equivalence. Whenever $\mu \wedge \xi, \nu \wedge \xi$ are regular sentences of $\mathscr{L}^\sharp$,*

Right Weakening: if $\mu \approx \nu$ and $\nu \models \xi$ then $\mu \approx \xi$.
Left Classical Equivalence: if $\mu \approx \nu$ and $\mu \equiv \xi$ then $\xi \approx \nu$.
Cautious Monotony: if $\mu \approx \nu$ and $\mu \approx \xi$ then $\mu \wedge \xi \approx \nu$.
Premiss Disjunction: if $\mu \approx \nu$ and $\xi \approx \nu$ then $\mu \vee \xi \approx \nu$.
Conclusion Conjunction: if $\mu \approx \nu$ and $\mu \approx \xi$ then $\mu \approx \nu \wedge \xi$.

The objective Bayesian semantics is sometimes preferred to the standard semantics on conceptual grounds (semantics in terms of rational degrees of belief is very natural, and objective Bayesianism provides a compelling account of rational degree of belief) and sometimes on pragmatic grounds—that a single probability rather than a set of probabilities often attaches to the conclusion of the fundamental question is computationally appealing and can simplify matters if a decision must made on the basis of an answer to such a question (Williamson, 2007b, §11).

Part II
Probabilistic Networks

Chapter 8
Credal and Bayesian Networks

In Part I we argued that the question of Schema (1.1), $\varphi_1^{X_1}, \ldots, \varphi_n^{X_n} \approx \psi^?$, provides a unifying framework for probabilistic logic, into which several important approaches to probabilistic inference slot. Now, in Part II, we turn to the problem of how one might answer this fundamental question.

We have seen that in many cases it suffices to restrict attention to convex sets of probability functions, and even, in the case of objective Bayesianism, often a single probability function. This restriction will be important in what follows, since it will allow us to exploit the computational machinery of probabilistic networks—in particular *credal networks* and *Bayesian networks*—to help us answer the fundamental question.

The task, then, is to find an appropriate Y such that $\varphi_1^{X_1}, \ldots, \varphi_n^{X_n} \approx \psi^Y$, where $X_1, \ldots, X_n, Y$ are *intervals* of probabilities. Our strategy is to use the left hand side, $\varphi_1^{X_1}, \ldots \varphi_n^{X_n}$, to determine a probabilistic network on the domain. This probabilistic network offers an efficient means of representing the probability functions that are models of the left hand side, and an efficient means of drawing inferences from these probability functions. One can then use this network to calculate Y, the range of probability values that models of the left hand side attach to ψ.

In §8.1 we introduce credal and Bayesian networks and the different configurations of these networks that will be required in Part II of this book. Then, in §8.2, we develop some common machinery for drawing inferences from probabilistic networks in order to answer our fundamental question. The remainder of Part II will be devoted to exploring the variety of networks produced by the different semantics of Part I. In each subsequent section we shall provide one or more algorithms for constructing a probabilistic network that suits the corresponding semantics. We will focus on explaining the basic steps of each construction algorithm in broad terms. Space constraints prohibit detailed analysis and justification of these algorithms but further details are available in the supporting references.

In summary, then, the recipe of this book is as follows:

- **Representation**: Formulate a question of the form of Schema (1.1).
- **Interpretation**: Decide upon appropriate semantics (Part I).

R. Haenni et al., *Probabilistic Logics and Probabilistic Networks*, Synthese Library 350, DOI 10.1007/978-94-007-0008-6_8, © Springer Science+Business Media B.V. 2011

- **Network Construction**: Construct a probabilistic network to represent the models of the left-hand side of Schema (1.1) (subsequent sections of Part II).
- **Inference**: Apply the common machinery of §8.2 to answer the question posed in the first step.

This book is by no means the first to advocate integrating probabilistic logics and probabilistic networks (see, e.g., Poole, 1993; Ngo and Haddawy, 1995; Jaeger, 2002; Kersting et al., 2006; Richardson and Domingos, 2006; Landwehr et al., 2007; Cozman et al., 2008; Laskey, 2008). Previous work has focussed on developing specific probabilistic logics that integrate probabilistic networks into the *representation scheme*—i.e., into the logical syntax itself. Here, however, we adopt a quite different approach, by using probabilistic networks as a *calculus* for probabilistic logic in general. Thus probabilistic networks play a role here analogous to that of proof in classical logic: they provide a means of answering the kind of question that the logic faces.

8.1 Kinds of Probabilistic Network

Suppose we have a finite propositional language $\mathcal{L}_V$ with propositional variables $V = \{A_1, \ldots, A_n\}$ which may take values *true* or *false*. As outlined in §1.5, the assignment $A_i = true$ is denoted by a_i^1 or simply a_i, while $A_i = false$ is denoted by a_i^0 or $\bar{a}_i$. A *Bayesian network* consists of a directed acyclic graph (DAG) on V, together with the probability functions $P(A_i|PAR_i)$ for each variable $A_i \in V$ conditional on its parents $PAR_i \subseteq V$ in the graph. (Here $P(a_i^e|par_i^e)$ refers to the probability of a particular assignment to variable A_i conditional on the probability of a particular assignment to its parents, while $P(A_i|PAR_i)$ refers to the probability function itself— i.e., the function that maps assignments to A_i and its parents to numbers in the unit interval.) An example of a Bayesian network is provided by the graph of Figure 8.1 together with corresponding conditional probability functions specified by:

$$P(a_1) = 0.1, \qquad P(a_2|a_1) = 0.9, \qquad P(a_3|a_2) = 0.4,$$
$$P(a_2|\bar{a}_1) = 0.3, \qquad P(a_3|\bar{a}_2) = 0.2.$$

Fig. 8.1: Directed acyclic graph in a Bayesian network.

The joint probability function P over $\mathcal{L}_V$ is assumed to satisfy the *Markov condition*: each variable is probabilistically independent of it's non-descendants ND_i conditional on its parents PAR_i, written $A_i \perp\!\!\!\perp ND_i|PAR_i$. Under this assumption, a

Bayesian network suffices to determine the joint probability function under the identity

$$P(A_1,\ldots,A_n) = \prod_{i=1}^{n} P(A_i|PAR_i). \tag{8.1}$$

In the above example,

$$P(a_1\bar{a}_2 a_3) = P(a_3|\bar{a}_2)P(\bar{a}_2|a_1)P(a_1) = 0.2 \times (1 - 0.9) \times 0.1 = 0.002.$$

A *credal network* consists of a directed acyclic graph on $V = \{A_1,\ldots,A_n\}$ together with closed convex sets $\mathbb{K}(A_i|PAR_i)$ of conditional probability functions $P(A_i|PAR_i)$, for each variable $A_i \in V$ conditional on its parents in the graph. As previously mentioned, a closed convex set of probability functions is called a *credal set*. (See §2.1.4 for more on convexity.) As an example of a credal network consider Figure 8.1 together with the local conditional distributions constrained by

$$P(a_1) \in [0.1, 0.2], \qquad P(a_2|a_1) = 0.9, \qquad P(a_3|a_2) \in [0.4, 1],$$
$$P(a_2|\bar{a}_1) \in [0.3, 0.4], \qquad P(a_3|\bar{a}_2) = 0.2.$$

8.1.1 Extensions

In a credal network, further independence assumptions are made in order to determine a joint credal set $\mathbb{K}(A_1,\ldots,A_n)$ over the whole domain from the given conditional credal sets. The credal network framework admits a number of choices as to which assumptions to make here.

Natural Extension. One choice is to make no independence assumptions at all. Then we have what in (Cozman, 2000) is called a *natural extension under no constraints*, i.e., the joint credal set

$$\mathbb{K}(A_1,\ldots,A_n) = \{P : P(A_i|PAR_i) \in \mathbb{K}(A_i|PAR_i)\},$$

whose elements all comply with the probabilistic constraints of the given conditional credal sets.

Strong Extension. Another possible choice is to take the convex hull of the *extremal points* defined by the local conditional credal sets and the conditional independencies determined from the graph via the Markov condition. The resulting joint credal set, is then called the *strong extension* of the credal network. It is noteworthy that not all probability assignments in the strong extension need comply with the independence relations suggested by the graph—only the extremal points need satisfy these contraints. This is further explained in §8.1.2 below.

Complete Extension. Yet another choice is to assume that *each* probability function in the joint credal set satisfies the Markov condition with respect to the given graph. In that case the credal network can be thought of as the set of Bayesian networks based on the given DAG whose conditional probabilities satisfy the constraints imposed by the conditional credal sets, $P(A_i|PAR_i) \in \mathbb{K}(A_i|PAR_i)$. We call the resulting credal set the *complete extension* of the credal network.[1]

Note that the natural extension coincides with the strong (or complete) extension with respect to a *complete* graph (a DAG in which each pair of variables is linked by an arrow, and which therefore implies no independencies via the Markov condition). Hence it suffices to consider strong and complete extensions, and we will not appeal to the concept of natural extension in what follows. (Our discussion of the standard semantics and Bayesian statistical inference appeals to complete extensions, while strong extensions apply to objective Bayesianism.) Note too that the extremal points are independent of extension (i.e., the extremal points of any one extension coincide with those of any other extension). In fact the inferences drawn from a credal network just depend on these extremal points. Our common inferential machinery will handle the extremal points and so may be applied to any of the above extensions.

8.1.2 Extensions and Coordinates

The notion of extension is tied up with the notion of convexity. In §2.1.4 we remarked that whether or not a set of probability functions is convex depends on the coordinate system used to specify the probability functions. In particular, depending on the choice of the coordinates for the space of probability functions, the strong extension of a network may be identical to the complete extension. We now explain and illustrate this with an example.

Example 12. Consider the credal network on variables A and B defined by the empty graph (no arrow between A and B) and local credal sets $P(a) \in [0.3, 0.7]$, $P(b) \in [0, 0.5]$. This credal network defines four extremal points, each representable by a Bayesian network on the empty graph (implying via the Markov condition that $A \perp\!\!\!\perp B$): one has local distributions $P(a) = 0.3$, $P(b) = 0$, a second has $P(a) = 0.7$, $P(b) = 0$, the third $P(a) = 0.3$, $P(b) = 0.5$ and the fourth $P(a) = 0.7$, $P(b) = 0.5$. The strong extension of this credal network is simply the convex hull of these four extremal points.

But now the question arises as to how the convex hull in the above example is defined. The convex hull of a set $\mathbb{P}$ of probability functions is the smallest set of probability functions that contains $\mathbb{P}$ and is closed under convex combination. A convex combination of functions P and Q is a function R such that $z = \lambda x + (1 - \lambda)y$ for each coordinate z of R, where x and y are the corresponding coordinates of P and Q respectively and $\lambda \in [0, 1]$.

One possibility is to specify a full joint distribution over the four atomic states $a \wedge b$, $a \wedge \neg b$, $\neg a \wedge b$, and $\neg a \wedge \neg b$, and take as coordinates $x_1 = P(a \wedge b)$, $x_2 =$

---

[1] A credal set is a closed convex set of probability functions. Convexity, however, is relative to coordinate system; the complete extension is convex with respect to the network-structure coordinate system. See §8.1.2 for further explanation.

$P(a \wedge \neg b)$, $x_3 = P(\neg a \wedge b)$, and $x_4 = P(\neg a \wedge \neg b)$. We represent these four mutually exclusive, exhaustive possibilities within a 3-dimensional unit simplex in Figure 8.2. Here the points where A and B are *not* dependent are those on the curved surface. The small quadilateral on the curved surface identifies the points where A and B are independent from one another *and* satisfy the given constraints on assignments.

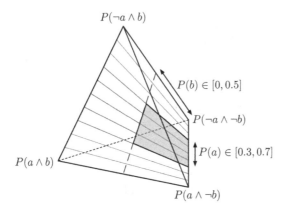

Fig. 8.2: Atomic-state coordinates.

This curved surface is also depicted on the left hand side of Figure 8.3. Note that it is not the case that all points in the convex hull of the small quadrilateral also lie on the curved surface: the dotted line on the left hand side of Figure 8.3 lies outside the surface. Hence probability functions within the strong extension fail to satisfy the independence relation $A \perp\!\!\!\perp B$ that is satisfied by the extremal points.

One can depict the full distribution over four possibilities using different coordinates. Now represent a probability function by coordinates $\alpha = P(a) = 1 - P(\bar{a})$, $\beta_1 = P(b|a) = 1 - P(\bar{b}|a)$, and $\beta_0 = P(b|\bar{a}) = 1 - P(\bar{b}|\bar{a})$. This coordinate system is depicted by the cube on the right hand side of Figure 8.3. Here A and B are independent, $A \perp\!\!\!\perp B$, iff $\beta_0 = \beta_1$. This is a linear restriction in the space of possible assignments to A and B—the diagonal plane in the cube, which corresponds to the curved surface in the coordinate system on the left hand side of Figure 8.3.

Again, the small quadrilateral represents the region satisfying the constraints imposed by the local credal sets. In this case, the convex hull of the extremal points of this region lies on the diagonal surface: all probability functions in the strong extension satisfy the independence relation imposed by the graph in the credal network, and hence the strong extension coincides with the complete extension.

This shows that the strong extension can depend on the coordinate system used to represent probability functions. Hence it is important when using a strong extension to specify the coordinate system. In probabilistic logic it is typical to use the former, atomic-state coordinate system (x_1, x_2, x_3, x_4). This ensures that if the extremal points all satisfy a linear constraint of the form $P(\varphi) \in X$ then so will every member of the strong extension. (For example, the objective Bayesian semantics of

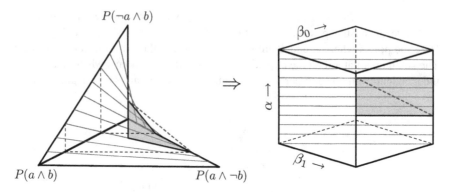

Fig. 8.3: Two different coordinate systems.

§7 appeals to the atomic-state coordinate system in determining the convex closure of the probability function that satisfy constraints imposed by evidence.) On the other hand, the latter, network-structure coordinate system $(\alpha, \beta_0, \beta_1)$ is often typical in discussions of causality, where it is more important that the strong extension preserve independence relations than linear constraints.

8.1.3 Parameterised Credal Networks

It is important to note that there are credal sets that cannot be represented by a credal network of the above form.

Example 13. Consider the credal set that consists of all probability functions on $V = \{A, B\}$ that satisfy the constraints $P(ab) = 0.3$ and $P(b|\bar{a}) = 0$. This implies $P(a) \in [0.3, 1]$, but the possible values of $P(b|a)$ depend on the particular value x of $P(a)$ in $[0.3, 1]$, since $P(b|a) = 0.3/x$. While it is true that $P(b|a) \in [0.3, 1]$, the value of $P(b|a)$ can not be chosen independently of that of $P(a)$. So the credal network defined on the graph with an arrow from A to B that has local credal sets $P(a) \in [0.3, 1]$, $P(b|a) \in [0.3, 1]$, and $P(b|\bar{a}) = 0$ does *not* represent the joint credal set. Note that the graph is complete so the choice of extension is irrelevant here.

To represent the credal set of the above example we need to appeal to what we call a *parameterised credal network*. The graph is as before but now the conditional credal sets involve a parameter: $x := P(a) \in [0.3, 1]$, $P(b|a) = 0.3/x$, and $P(b|\bar{a}) = 0$. By invoking the parameter x, one avoids the assumption that the local credal sets operate independently as constraints on P. Parameterised credal networks can be used to represent arbitrary credal sets.

8.2 Algorithms for Probabilistic Networks

There has been a tremendous amount of research directed at finding the most efficient algorithms for inference in Bayesian and credal networks. Nearly all cutting-edge algorithms for Bayesian networks use some sort of *local computation* techniques (Lauritzen and Spiegelhalter, 1988; Shenoy and Shafer, 1988; Kohlas and Shenoy, 2000), which try to systematically bypass operations on large state spaces. In the worst-case, these algorithms run in time and space exponential to the network's *induced treewidth*, which is an indicator of the 'density' of the available independence assumptions. Further improvements are obtained by exploiting so-called *context-specific independencies* (Boutilier et al., 1996; Wachter et al., 2007). These techniques deal with local independence relations within (rather than between) the given conditional probability tables.

In the case of credal networks, finding efficient inference algorithms is much more challenging. There are some attempts to apply local computation techniques to credal networks, but the problem remains unfeasible for most non-trivial instances. Compared to Bayesian networks, the additional computational complexity results from the potentially unbounded number of extremal points needed to describe arbitrary credal sets, which can quickly undermine any benefits of local computation. In fact, one can see inference in credal networks as a global multilinear optimization problem on top of the given network structure (Cozman and de Campos, 2004), which is intractable on its own. Facing this inherent computational complexity, exact inference methods are only exceptionally applicable to credal networks, and this is the reason why most of the current research in this area aims at approximating the results rather than computing them precisely—see (Cano and Moral, 1996; Cano et al., 2002; da Rocha and Cozman, 2003; da Rocha et al., 2003; Ide and Cozman, 2004, 2005; Antonucci et al., 2006; Haenni, 2007). What is common to those methods is a general distinction between *inner* and *outer approximations*, depending on whether the approximated interval is enclosed in the exact solution or vice versa.

In the remainder of this section, our goal is to single out an algorithm for probabilistic networks which satisfies the particular requirements imposed by the framework of this book. In the following sections, we shall then explore the network construction problem for each of the proposed semantics.

8.2.1 Requirements of the Probabilistic Logic Framework

Despite the huge number of available algorithms for probabilistic networks, only a few are valuable candidates for solving problems of the form of Schema (1.1). One reason for this is the fact that nearly all proposed semantics for probabilistic logic deal with sets of probabilities and are thus inherently tied to credal networks. In other words, inference algorithms for Bayesian networks are not general enough to solve instances of Schema (1.1). Another reason hinges on the following particular requirements, which are usually not of primary concern when building algorithms for probabilistic networks:

- Given that the conclusion ψ on the right hand side of Schema (1.1) may be an arbitrary logical formula, the algorithm must be able to compute probabilities $P(\psi)$ or respective intervals $[\underline{P}(\psi), \overline{P}(\psi)]$ of arbitrarily complex queries. If ψ is a propositional formula, a general strategy to handle such situations is to decompose ψ into pairwise disjoint terms, thus producing what is called a *disjoint sum-of-products*. This can be done in two steps: first transform ψ into a disjunctive normal form (by applying de Morgan's laws), and then apply Abraham's algorithm to obtain $\psi \equiv \psi_1 \vee \cdots \vee \psi_r$, where each ψ_i is a conjunction of literals (term) satisfying $\psi_i \wedge \psi_j \equiv \bot$ for all $j \neq i$ (Abraham, 1979; Anrig, 2000). This particular representation of ψ can then be used to compute $P(\psi) = \sum_{i=1}^{r} P(\psi_i)$ as a sum of probabilities $P(\psi_i)$, each of them involving a simple term ψ_i only. The problem of computing probabilities of complex queries is thus reducible to the problem of computing term probabilities, and the problem could be solved by querying the network r times, once for each individual term. But what we would ideally expect from an algorithm is the ability to compute such term probabilities simultaneously.

- Given that the premises on the left hand side of Schema (1.1) may include functionally interrelated probability intervals, e.g., to express conditional independencies (see §2.2), we would expect an algorithm to cope with the parameters involved in the description of those interrelated intervals. In other words, what we need is a general machinery for parameterised credal networks. Following the remarks of §8.1.3, we assume linearity with respect to the chosen coordinate system, in order to ensure that extremal points are easily identifiable.

The method sketched in the remainder of this section is a candidate for common inferential machinery for various probabilistic logics. The core of the method has originally been proposed as an exact inference algorithm for Bayesian networks (Darwiche, 2002; Chavira and Darwiche, 2005, 2007; Wachter and Haenni, 2007), but it is extensible to approximate inference in credal networks (Haenni, 2007). As we will see, it can also be used to simultaneously deal with multiple queries and to handle functionally interrelated probability intervals, i.e., it meets all the above requirements.

8.2.2 Compiling Probabilistic Networks

The general idea of compiling a probabilistic network is to split up inference into an expensive *compilation phase* and a cheap *query answering phase*. For a given network, the foregoing compilation phase is a unique step, which may take place well before the actual inferences need to be conducted. Computationally, the compilation may be very expensive, but once the compiled network is available, it is guaranteed that all sorts of probabilistic queries can be answered efficiently.

To implement this simple idea, the first step is to represent the topological structure of a probabilistic network by propositional sentences. The resulting propositional encoding of the network is then transformed into a generic graphical representation called *deterministic, decomposable negation normal form*, or simply d-DNNF (Darwiche and Marquis, 2002; Wachter and Haenni, 2006b). One can think of a d-DNNF as a directed acyclic graph whose internal nodes represent logical AND's and OR's, while the leaves at the bottom of the graph represent literals in a propositional language. The internal nodes are such that the children of an AND-node do not share common variables (called *decomposability*) and the children of an OR-node are pairwise logically inconsistent (called *determinism*). These properties are crucial to conduct probability computations.

The transformation from the original network encoding into a generic d-DNNF is a unique step called *compilation*, the computationally hardest task of the whole procedure. It can be implemented in various ways, e.g., by a method called *tabular compilation* (Chavira and Darwiche, 2007), which is essentially a variable elimination procedure that exploits the benefits of local computations. Therefore, it runs in time and space exponential to the network's induced treewidth, similar to standard inference algorithms for Bayesian networks. A compiled Bayesian network can then be used to answer arbitrary probabilistic queries in time linear to its size. The computational task is thus divided into an expensive (off-line) compilation phase and an efficient (on-line) query-answering phase. The latter is further illustrated by the following example.

Example 14. Consider a probabilistic network with two propositional variables A and B and an arrow pointing from A to B. The specification of the network requires thus three parameters for $P(a)$, $P(b|a)$, and $P(b|\bar{a})$. Consider for each of these parameters an auxiliary propositional variable and let $\{\theta_a, \theta_{\bar{a}}\}$, $\{\theta_{b|a}, \theta_{\bar{b}|a}\}$, and $\{\theta_{b|\bar{a}}, \theta_{\bar{b}|\bar{a}}\}$ be their respective sets of values. These values are used as literals at the bottom of the d-DNNF obtained from compiling the network. The result of the compilation is shown on the left hand side of Figure 8.4.

Now let's suppose we want to use the compiled network to compute $P(b)$. This involves three steps. First, we need to instantiate the d-DNNF for the query b, which essentially means to set the literal b to *true* and the literal $\bar{b}$ to *false*. This allows a number of simplifications, resulting in the d-DNNF shown on the right hand side of Figure 8.4 (upper part). Second, we need to eliminate the variables not included in the query, in our case the variable A. For this, we can set both literals a and $\bar{a}$ to *true*, which again allows a number of simplifications.[2] The result of these simplifications is shown on the right hand side of Figure 8.4 (lower part). Finally, we need to propagate concrete values for the parameters in a simple bottom-up procedure. For this, each AND-node multiplies and each OR-nodes sums up the values of its children.

---

[2] In general, it is infeasible to eliminate variables from a d-DNNF, but in the particular case of a compiled probabilistic network, it turns out that the variables to be eliminated are always such that eliminating them is as simple as instantiating the query variables (Wachter and Haenni, 2006a, 2007).

In our example, this yields $P(b) = P(a)P(b|a) + P(\bar{a})P(b|\bar{a})$, which is exactly what we would expect.

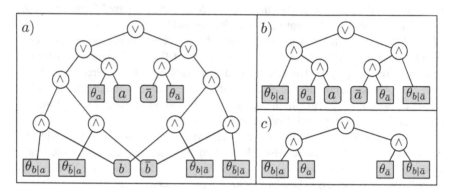

Fig. 8.4: Querying a compiled probabilistic network: (a) The compiled network. (b) The compiled network after instantiating the query b. (c) The instantiated compiled network after eliminating A.

The procedure illustrated in Example 14 may seem a bit cumbersome to solve such simple examples, but the same scheme is generally applicable to arbitrarily complex probabilistic networks and always runs in time linear to the size of the compiled d-DNNF structure. This is also true if the query is a conjunction of literals and for networks with multi-state variables (Wachter and Haenni, 2007). Moreover, by postponing the numerical computations to the very last step of the procedure, we obtain a very flexible way of updating the computations when some of the numerical parameters change. We will show next how to exploit this flexibility to approximate inference in credal networks.

8.2.3 The Hill-Climbing Algorithm for Credal Networks

In a credal network, each network parameter is tied to an interval rather than a sharp value. This defines a credal set over the involved variables and it means that we get an infinity of choices when it comes to propagating the numerical values in the resulting d-DNNF. Fortunately, we know that in order to get lower and upper bounds for our query, it is sufficient to work with the extremal points of the involved credal sets (see §8.1.1). In other words, we can restrict the infinitely large search space to a finite search space which involves all possible combinations of lower and upper bounds of the given intervals. However, as the number of such combinations grows exponentially with the number of parameters and thus with the size of the network, it is impracticable to conduct an exhaustive search for reasonably large networks.

Example 15. Consider the result for the query $P(b)$ in the previous example: $P(b) = P(a)P(b|a) + P(\bar{a})P(b|\bar{a})$. Now suppose that $P(a) \in [0.2, 0.5]$, $P(b|a) \in [0.3, 0.4]$, and $P(b|\bar{a}) \in [0.6, 0.9]$ are the constraints for the three network parameters. The search space to find the lower and upper bounds for $P(b) \in \mathbb{K}(b)$ is therefore of size $2^3 = 8$. The corresponding extremal points are listed in Table 8.1, from which we conclude that $\underline{P}(b) = 0.45$ and $\overline{P}(b) = 0.8$ are the required lower and upper bounds, respectively.

Table 8.1: The Eight Extremal Points of the Convex Set $\mathbb{K}(A,B)$ in the Credal Network of the Current Example

$P(a)$	$P(\bar{a})$	$P(b\|a)$	$P(b\|\bar{a})$		$P(b)$
		0.3	0.6	$\Rightarrow$	0.54
			0.9	$\Rightarrow$	0.78
0.2	0.8				
		0.4	0.6	$\Rightarrow$	0.56
			0.9	$\Rightarrow$	$0.80 = \overline{P}(b)$
		0.3	0.6	$\Rightarrow$	$0.45 = \underline{P}(b)$
			0.9	$\Rightarrow$	0.60
0.5	0.5				
		0.4	0.6	$\Rightarrow$	0.50
			0.9	$\Rightarrow$	0.65

The fact that an exhaustive search is too expensive in general requires us to perform some sort of approximation. A generic combinatorial optimization technique, which is widely used in similar AI-related applications, is called *hill-climbing* (Russell and Norvig, 2003). The goal of hill-climbing is to maximize (or minimize) a function $f : X \to \mathbb{R}$ through local search, where X is usually a discrete multi-dimensional state space. Local search means jumping from one configuration in the state space to a neighboring one, until a local or possibly the global maximum (or minimum) is reached. This step is usually iterated for some time with randomly generated starting points, thus making it an interruptible-anytime algorithm.

To approximate lower and upper bounds when querying a credal network, we can apply the hill-climbing approach to the search space defined by the credal set's extremal points. One can think of it as a walk along the edges of the credal set's convex hull. Using the resulting d-DNNF after compiling and querying the network, we can easily update the numerical result for the current extremal point. As demonstrated in (Haenni, 2007), the d-DNNF also supports the determination of the neighbour with the steepest ascent (or descent), which is important for optimizing the performance of the hill-climbing procedure.

Example 16. Consider the credal network from Example 14 and the list of extremal points from Table 8.1, and suppose we first want to compute the upper bound $\overline{P}(b)$. Let us start the hill-climbing by first considering the lower bounds of all given intervals, i.e., $P(a) = 0.2$, $P(b|a) = 0.3$, and $P(b|\bar{a}) = 0.6$, which yields $P(b) = 0.54$.

Then suppose we decide to change $P(b|\bar{a})$ from 0.6 to 0.9 to get $P(b) = 0.78$, which is a first major improvement. If we then decide to change $P(b|a)$ from 0.3 to 0.4, we have already found the global maximum $\overline{P}(b) = 0.8$ after the first hill-climbing iteration. For the lower bound $\underline{P}(b)$, let us start from scratch with all lower bounds. By simply changing $P(a)$ from 0.2 to 0.5, we immediately find the global minimum $\underline{P}(b) = 0.45$. Note that it is the simplicity of the example that prevents us from running into a local maximum or minimum. The whole search space of extremal points is shown on the left hand side of Figure 8.5, where each arrows indicatesa an uphill path between two neighbors.

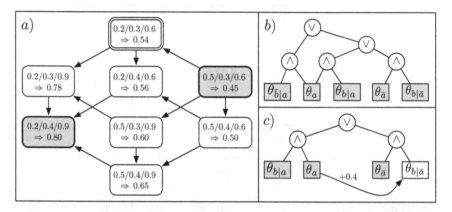

Fig. 8.5: (a) Hill-climbing in the search space of extremal points. (b) The compiled network instantiated for $a \vee b$. (c) Querying a parameterized credal network.

For more details on this procedure and for a formal description of the hill-climbing algorithm we refer to (Haenni, 2007).

8.2.4 Complex Queries and Parameterised Credal Networks

To conclude this section, let us discuss how to use the proposed hill-climbing algorithm to meet the particular requirements discussed earlier in this section. The first point was to allow arbitrarily complex queries, which can be solved by querying the network for a corresponding set of disjoint terms. This means that the compiled network needs to be instantiated individually for each term. The result of this is a collection of d-DNNFs, which are likely to overlap heavily for large networks. Such a structure with multiple roots is called *shared d-DNNF*, and it supports the same bottom-up propagation to compute probabilities. We can thus conduct the same hill-climbing procedure, except that we need to maximize (or minimize) the sum of the values obtained at each root. Or we may simply connect all individual roots with an

OR-node and then maximize (or minimize) the value obtained at this newly created root.

Example 17. Consider the same credal network from the previous examples, but now let $\psi = a \vee b$ be the hypothesis of interest. The first step is to write ψ as a disjoint sum-of-products $(a \wedge \neg b) \vee b$, which is obviously equivalent to $a \vee b$. We need thus to instantiate the compiled network twice, once for $a \wedge \neg b$ and once for b. For $a \wedge \neg b$ we simply get an AND-node which connects θ_a and $\theta_{\bar{b}|a}$, and for b the result is as discussed before. The result of connecting them by an OR-node is depicted on the right hand side of Figure 8.5 (upper part). This is the structure on which the hill-climbing procedure operates for the query $a \vee b$.

The second particular requirement of the probabilistic logic framework is the ability to deal with parameterised credal networks. This means that we may need to take care of intervals that are functionally interrelated. In the proposed hill-climbing procedure, this is quite easy to realize: we only need to make sure that the set of selected extremal points is always compatible with the given functional dependencies. This makes the transition from one extremal point to a neighboring one slightly more complicated, but in return we may get a corresponding search space reduction.

Example 18. Consider the credal network from Example 15 and let $x := P(a) \in [0.2, 0.5]$. Now suppose that $P(b|\bar{a}) \in [0.6, 0.9]$ is replaced by $P(b|\bar{a}) = x + 0.4$, i.e., $P(b|\bar{a})$ functionally depends on $P(a)$ as indicated on the right hand side of Figure 8.5 (lower part). This reduces the number of network parameters from 3 to 2 and therefore the size of the search space from $2^3 = 8$ to $2^2 = 4$. In the hill-climbing procedure discussed in Example 16, this reduction prevents us from jumping from the initial result $P(b) = 0.54$ to $P(b) = 0.78$ by changing the value of $P(b|\bar{a})$. Instead we may first change $P(b|a)$ from 0.3 to 0.4 to get $P(b) = 0.56$. If $P(a)$ is then changed from 0.2 to 0.5, we must also change $P(b|\bar{a})$ from 0.6 to 0.9, and this leads to the new upper bound $\overline{P}(b) = 0.65$. Similarly, we get a new lower bound $\underline{P}(b) = 0.54$.

Chapter 9
Networks for the Standard Semantics

In §2 we introduced the standard semantics for the inferential problem expressed in Schema (1.1). In this section we investigate how credal networks can be employed to improve the reach and efficiency of inferences in the standard semantics. After that we briefly discuss some conceptual issues that arise from inference in a credal network.

9.1 The Poverty of Standard Semantics

In the standard semantics, the inferences run from credal sets, determined by constraints on the probability assignments, to other credal sets. Clearly, the probability assignments in the credal sets are defined by the axioms of Kolmogorov. These axioms function as basic rules of inference, in the sense that they allow us to relate constraints in the premises to constraints in the conclusion. As we made clear in §2, all the constraints on the probability assignments derive from explicit premises. The standard semantics on itself imposes no further constraints.

In §8 we explained that a credal set can be summarised conveniently in a credal network, and that such networks can be used to speed up the inference process. But it is only when we assume the so-called strong or complete extension of a credal network that we can fully employ their computational and conceptual advantages. Moreover, taking the strong or complete extension of a credal network is only really useful if some edges are missing from the network. In this respect a credal network representation works the same as a Bayesian network representation: the advantages of independence and local computation are most salient for sparse networks.

As also explained in the foregoing, taking the strong or complete extension of an incomplete graph amounts to making specific independence assumptions. The idea is to use the specifics of the domain of application, as captured by the semantics of the inference, to motivate these assumptions. However, the standard semantics does not have any specific domain of application, and so the additional independence assumptions cannot be motivated from semantic considerations. Therefore, in the

R. Haenni et al., *Probabilistic Logics and Probabilistic Networks*, Synthese Library 350, DOI 10.1007/978-94-007-0008-6_9, © Springer Science+Business Media B.V. 2011

standard semantics credal networks are only useful if the premises themselves include probabilistic independence assumptions.

9.2 Constructing a Credal Net

Suppose that we want to deal with an inference problem in the standard semantics, and that the premises include certain explicit probabilistic independencies. Let I be the set of these probabilistic independencies. One can then construct a credal network (under a complete extension) as follows.

Algorithm 5 Construction of a credal network for the standard semantics, based on explicit independence assumptions.

Input: a set $V = \{A_1, \ldots, A_M\}$ of propositional variables, a set I of probabilistic independence relations of the form $A_i \perp\!\!\!\perp A_j$ or $A_i \perp\!\!\!\perp A_j | A_k$, and premises $\varphi_1^{X_1}, \ldots, \varphi_N^{X_N}$ involving those variables.

1. **Construction of a graph:** application of the adapted PC-algorithm of Pearl (Pearl, 2000) to find the smallest network $\mathcal{G}$ that satisfies the set of independence relations I.

   a. Start off with a complete undirected graph on V.
   b. For $n = 0, 1, 2, \ldots$, remove any edges A_i—A_j if $A_i \perp\!\!\!\perp A_j | X$ for some set X of n neighbours of A_i.
   c. For each structure A_i—A_j—A_k in the graph with A_i and A_k not adjacent, substitute $A_i \longrightarrow A_j \longleftarrow A_k$ if A_j was not found to screen off A_i and A_k in the previous step.
   d. Repeatedly substitute:
      i. $A_i \longrightarrow A_j \to A_k$ for $A_i \to A_j$—A_k with A_i and A_k non-adjacent;
      ii. $A_i \longrightarrow A_j$ for A_i—A_j if there is a chain of arrows from A_i to A_j;
      iii. $A_i \longrightarrow A_j$ for A_i—A_j if there are two chains A_i—$A_k \longrightarrow A_j$ and A_i—$A_l \longrightarrow A_j$ with A_k and A_l not adjacent;
      iv. $A_i \longrightarrow A_j$ for A_i—A_j if there is a chain A_i—$A_k \longrightarrow A_l \longrightarrow A_j$ with A_k and A_j not adjacent.

Finally, define the network coordinates $P\left(a_k^1 | a_j^{c(j)} \cdots a_{j'}^{c(j')}\right) = \gamma_{k|j\cdots j'}^{c(j)\cdots c(j')}$, or γ_k for short.

2. **Derivation of constraints on network coordinates:** deriving the restrictions to the conditional probability assignments in the network $\mathcal{G}$, along the following lines.

   a. For each premise $\varphi_i^{X_i}$, collect the rank number of the propositional variables A_j that appear in φ_i, and all those propositional variables $A_{j'}$ that are parents of these A_j, in the vector s_i, whose elements are denoted $s_i(n)$. Order them such that $s_i(n-1) < s_i(n)$.

b. Rewrite the expression φ_i in a disjunctive normal form of the propositional variables $A_{s_i(n)}$ according to $\bigvee_m \bigwedge_n A_{s_i(n)}^{e_i(m,n)}$. In this expression $e_i(m,n) \in \{0,1\}$ are the elements in the matrix e_i whose rows $e_i(m)$ indicate whether the m-th conjunction in the disjunctive form contains $A_{s_i(n)}^0$ or $A_{s_i(n)}^1$.

c. Compute the probability $P\left(\bigwedge_n A_{s_i(n)}^{e_i(m,n)}\right)$ in terms of the network coordinates, by relying on the valuations of propositions of lower rank. We have

$$\Gamma_i^m = \prod_n \gamma_{s_i(n)|s_i(1)\cdots s_i(n-1)}^{e_i(m,1)\cdots e_i(m,n-1)}.$$

Finally, compute the probability

$$P(\varphi_i) = P\left(\bigvee_m \bigwedge_n A_{s_i(n)}^{e_i(m,n)}\right) = \sum_m \Gamma_i^m.$$

d. Impose the constraints $P(\varphi_i) = \sum_m \Gamma_i^m \in X_i$, and collect them in a single constraint set Γ for all the φ_i. This constraint set is a system of linear equations in the coordinates $\gamma_{k|j\cdots j'}^{c(j)\cdots c(j')}$, or $\gamma_k^{c_k}$ for short.

e. Compute the extremal points v_i for the coordinates involved in the equations Γ. They can be found by iteratively solving Γ for the upper and lower bounds on them. We start at $i = M = 1$.

  i. Index the coordinates involved in the constraints with m. In the extremal point indexed i, $v_i(m)$ denotes the value of coordinate m in the point indexed i.

  ii. If there are unique solutions for Γ among the free coordinates in v_i, set the corresponding entries in $v_i(m)$ to these solutions.

  iii. If all the entries in $v_i(m)$ have a value, start on a new extremal point, setting $i = i+1$. If $i > M$, then go to the last line: all the extremal points have been found.

  iv. Let n be the number of free parameters in v_i; for convenience we number them with $j = 1, \ldots, n$. For i' from M down to $i+1$, copy $v_{i'+2n-1} = v_{i'}$ and erase $v_{i'}$. If $i = M$, skip this step.

  v. Set $M := M + 2n - 1$. We have made a new set of extremal points v_{i+1} to v_{i+2n-1}.

  vi. Now for i' from i to $i+2n-1$, set $v_{i'}(m) = v_i(m)$ for all $v_i(m)$ that have a fixed value. Calculate the interval $[l_m, u_m]$ for each $v_i(m)$ whose value is still undetermined, and set

$$v_{i+2(j-1)}(m) = l_m \qquad v_{i+2j-1}(m) = u_m.$$

  Each new extremal point is thus given one additional extremal value for one of the free parameters.

  vii. Now return to step ii. to solve the remaining free coordinates in the vector v_i.

  viii. Check for duplicates in the set of extremal points v_i and remove them.

Output: a graph $\mathcal{G}$ and a set of extremal points S in the network coordinates γ_k corresponding to the independence relations I and the premises $\varphi_1^{X_1}, \ldots, \varphi_n^{X_n}$.

The determination of the extremal points can be laborious. If in total there are $\dim(\Gamma) = N$ coordinates involved in the constraints, there are at most $N!2^N$ extremal points. But generally the number of points will be much smaller. To see what the algorithm comes down to in practice, consider the following example.

Example 19. Let A_j for $j = 1, 2, 3$ be the propositional variables V, and let $I = \{A_2 \perp\!\!\!\perp A_3 \mid A_1\}$ be the set of independence relations. Pearl's adapted PC-algorithm gives us the graph shown in Figure 9.1. The network coordinates are $P(a_1) = \gamma_1$, $P(a_2 \mid a_1^i) = \gamma_{2|1}^j$, and $P(a_3 \mid a_1^i) = \gamma_{3|1}^j$.

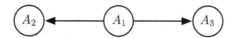

Fig. 9.1: The graph expressing that $A_2 \perp\!\!\!\perp A_3 \mid A_1$.

Now imagine that we have the following constraints on the variables in the graph: $\varphi_1^{X_1} = ((a_1 \vee \neg a_2) \rightarrow a_3)^{[1/3, 2/3]}$ and $\varphi_2^{X_2} = (a_3)^{1/2}$. We then apply the algorithm for the derivation of constraints on the coordinates γ_k as follows.

- We collect and order the rank numbers of the propositional variables appearing in the premises in two vectors $s_1 = \langle 1, 2, 3 \rangle$ and $s_2 = \langle 1, 3 \rangle$.
- We rewrite the expression $\varphi_1 = (a_1 \vee \neg a_2) \rightarrow a_3$ in a disjunctive normal form, and compute the probabilities in terms of the network coordinates,

Disjunct	Vector e_1	Probability Γ_1^m
$\neg a_1 \wedge \neg a_2 \wedge \neg a_3$ $\;e_1(1) = \langle 0,0,0 \rangle$		$(1 - \gamma_1)\left(1 - \gamma_{2|1}^0\right)\left(1 - \gamma_{3|1}^0\right)$
$a_1 \wedge \neg a_2 \wedge \neg a_3$ $\;e_1(2) = \langle 1,0,0 \rangle$		$\gamma_1 \left(1 - \gamma_{2|1}^1\right)\left(1 - \gamma_{3|1}^1\right)$
$a_1 \wedge a_2 \wedge \neg a_3$ $\;e_1(3) = \langle 1,1,0 \rangle$		$\gamma_1 \gamma_{2|1}^1 \left(1 - \gamma_{3|1}^1\right)$
$a_1 \wedge \neg a_2 \wedge a_3$ $\;e_1(4) = \langle 1,0,1 \rangle$		$\gamma_1 \left(1 - \gamma_{2|1}^1\right)\gamma_{3|1}^1$
$a_1 \wedge a_2 \wedge a_3$ $\;e_1(5) = \langle 1,1,1 \rangle$		$\gamma_1 \gamma_{2|1}^1 \gamma_{3|1}^1$
$(a_1 \vee \neg a_2) \rightarrow a_3$		$(1 - \gamma_1)\left(1 - \gamma_{2|1}^0\right)\left(1 - \gamma_{3|1}^0\right) + \gamma_1$

and similarly for a_3,

Disjunct	Vector e_2	Probability Γ_2^m
$\neg a_1 \wedge \neg a_2 \wedge a_3$ $\;e_2(1) = \langle 0,0,1 \rangle$		$(1 - \gamma_1)\left(1 - \gamma_{2|1}^0\right)\gamma_{3|1}^0$
$a_1 \wedge \neg a_2 \wedge a_3$ $\;e_2(2) = \langle 1,0,1 \rangle$		$\gamma_1 \left(1 - \gamma_{2|1}^1\right)\gamma_{3|1}^1$
$\neg a_1 \wedge a_2 \wedge \neg a_3$ $\;e_2(3) = \langle 0,1,1 \rangle$		$(1 - \gamma_1)\gamma_{2|1}^0 \gamma_{3|1}^1$
$a_1 \wedge \neg a_2 \wedge a_3$ $\;e_2(4) = \langle 1,1,1 \rangle$		$\gamma_1 \gamma_{2|1}^1 \gamma_{3|1}^1$
a_3		$(1 - \gamma_1)\gamma_{3|1}^0 + \gamma_1 \gamma_{3|1}^1$

- Impose the constraints $P(\varphi_i) = \sum_m \Gamma_i^m \in X_i$, collecting them in the following system of linear equations:

$$(1 - \gamma_1)\left(1 - \gamma_{2|1}^0\right)\left(1 - \gamma_{3|1}^0\right) + \gamma_1 \geq 1/3$$

$$(1 - \gamma_1)\left(1 - \gamma_{2|1}^0\right)\left(1 - \gamma_{3|1}^0\right) + \gamma_1 \leq 2/3$$

$$(1 - \gamma_1)\gamma_{3|1}^0 + \gamma_1\gamma_{3|1}^1 = 1/2.$$

- We compute the extremal points by iteratively filling in extremal values of each of these coordinates, according to the algorithm set up in the foregoing. To keep this reasonably brief, we will only deal with the first couple of vectors. For convenience we write $\gamma_2^0 = \gamma_2$, $\gamma_3^0 = \gamma_3$, and $\gamma_3^1 = \gamma_4$.

  - There are no unique solutions from the constraints, so there are $n = 4$ free coordinates. We create 7 new vectors v_i, and we set M to 8. There were no vectors $v_{i'}$ for $i' > i$ yet, so there is no need to copy any of them downwards.
  - We compute the bounds $0 \leq \gamma_1 \leq 2/3$, and set $v_1(1) = 0$ and $v_2(1) = 2/3$. We compute similar bounds for γ_k with $k = 2, 3, 4$, and fill in v_3 to v_8.
  - We return to line (ii) and check whether we can solve Γ under the additional constraint that $\gamma_1 = 0$. And we can: we fill in $\gamma_3 = 1/2$.
  - We now have $n = 2$ free parameters left. So we copy vectors v_2 to v_8 down into v_5 to v_{11}, and create the empty lines v_2 to v_4. Then for $i' = 2, 3, 4$ we copy $v_{i'}(1) = v_1(1) = 0$ and $v_{i'}(3) = v_1(3) = 1/2$ into the new lines.
  - We then compute the bounds $0 \leq \gamma_2 \leq 1/3$ and $0 \leq \gamma_4 \leq 1$, and fill them in: $v_1(2) = 0$, $v_2(2) = 1/3$, $v_3(4) = 0$, and $v_4(4) = 1$.
  - We return to line (ii) and see if we can solve the last remaining free coordinate in vector v_1. We cannot, so we create one copy of v_1 in v_2, copy all other v_i downwards, compute the bounds $0 \leq \gamma_4 \leq 1$, and fill in $v_1(4) = 0$ and $v_2(4) = 1$, thus completing the computation of the first two extremal points.
  - It is readily seen that extremal point v_3, which has $v_3(2) = 1/3$, has the same bounds $0 \leq \gamma_4 \leq 1$, so that we again copy all further v_i downwards, and fill in $v_3(4) = 0$ and $v_4(4) = 1$.
  - We deal with the next four extremal points, that have $v_i(1) = 0$ and $v_i(2) = 1/3$, in similar fashion. The difference is that we have here filled in $v_i(4)$ already, and we compute upper and lower bounds for γ_2, filling in $v_i(2)$. But the results are pairwise identical to the first four extremal points.
  - We now start on v_9, which has $v_9(1) = 2/3 \ldots$

- Once all extremal points have been found, the algorithm halts. We output the graph depicted above and a set of points v_i.

As indicated, the derivation of constraints on the coordinates of the network is computationally costly, especially for dense graphs. However, it must be remembered that after the construction of the graph and the derivation of constraints, any further inference is relatively fast. As before, the compilation of the network can be seen as an investment into future inferences: once the network is built, querying it is fast and easy.

9.3 Dilation and Independence

The use of credal sets leads to a rather awkward result when it comes to updating the probability assignments to new information by conditionalisation. The probability interval may get, what is called, *dilated*; see (Seidenfeld and Wasserman, 1993). We briefly discuss dilation here because it relates to the independence assumptions needed for using the strong or complete extension of a credal network.

Consider the credal set given by $P(a) = 1/2$ and $P(b) = 1/2$, but without an independence relation between A and B. In terms of the coordinates $\langle \alpha, \beta_0, \beta_1 \rangle$ used in §8.1.2, we can write $P(a) = \alpha = 1/2$ and $P(b) = \alpha\beta_1 + (1 - \alpha)\beta_0 = 1/2$, so that the credal set is defined by $\alpha = 1/2$ and $\beta_1 = 1 - \beta_0$. This set is represented by the thick line in Figure 9.2.

Now imagine that we decide to observe whether a or $\bar{a}$. Paradoxically, this decision is sufficient to throw us into doubt over the value of the probability of b, no matter what the result of the observation will be! From the probability assignments it follows that, whether we learn a or $\bar{a}$, the probability for b changes from a sharp value to the whole interval after the update. This is because when we learn a and condition on it, the probability for b becomes $P(b|a) = \beta_1 \in [0,1]$, and similarly, when we learn $\bar{a}$ and condition on it, it becomes $P(b|\bar{a}) = \beta_0 \in [0,1]$, as can be read off from Figure 9.2. We say that the probability for b is dilated by the partition $\{a, \bar{a}\}$.

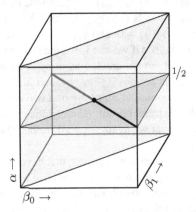

Fig. 9.2: The credal set represented in the coordinate system $\langle \alpha, \beta_0, \beta_1 \rangle$, as the thick line in the square $\alpha = 1/2$. For each of the points in this credal set we have $P(b) = 1/2$, but both β_0 and β_1 can take any value in $[0,1]$. The credal set associated with independence between A and B is depicted as a shaded area, $\beta_0 = \beta_1$. By the addition of the independence assumption, the credal set becomes the intersection of the thick line and the independence area, namely the singleton $\{\langle 1/2, 1/2, 1/2 \rangle\}$.

This awkward situation comes about because the restrictions on the probability assignment do not tell us anything about the relations between A and B. Accordingly, dilation is avoided if we assume an independence relation between A and B. As illustrated in Figure 9.2, if we add the independence constraint that $\beta_0 = \beta_1$, then the credal set is confined to $\alpha = \beta_0 = \beta_1 = 1/2$. Learning whether a or $\bar{a}$ then leaves the probability of b invariant.

This resolution ties in nicely with the philosophical discussion on dilation. Dilation is paradoxical, exactly because seemingly irrelevant information can introduce new uncertainty into an epistemic state. Say that we are at risk of developing a severe medical condition, which is assigned a probability half. This is bad news, but at least we know what we are in for. Incidentally, we do not yet know what weather it is going to be tomorrow, rain or sun, so we assign to both these events a probability of half as well. Upon waking up we see a drizzle outside, and in distress we call our doctor, because we are thrown into complete ignorance over whether we will develop the severe condition. Moreover, after a little reflection, we realise that sunshine had not meant much else for us!

It seems that we are right to call our doctor, indeed, but only because we have lost the plot. For one, it is confused that we are thrown into doubt independently of what the weather turns out to be. However, the root of the puzzle is that we do not consider the physical condition independent from the weather. So our beliefs, as expressed by the credal set, are flawed. If instead we represent our beliefs with a credal set consisting of probability functions that satisfy the independence of weather and health condition, we build the independence in from scratch. In that case, as illustrated above, we avoid the paradoxical conclusions.

Chapter 10
Networks for Probabilistic Argumentation

In a classical probabilistic argumentation system, in which the available information is encoded by a set of logical sentences Φ and a probability measure P over a subset of variables W appearing in Φ (see Part I, §2), we may already use a probabilistic network for the specification of P. In the simplest case, in which we use a fully specified Bayesian network to obtain a unique P, this leads to a probabilistic argumentation system according to Definition 4, and we may then compute sharp degrees of support and possibility for our hypotheses of interest. As an example of how to enhance a Bayesian network by a set of logical constraints, and of how to use the enhanced Bayesian network for reasoning and inference, this simple setting is interesting in its own right.

10.1 Probabilistic Argumentation with Credal Sets

If we depart from the assumption of a unique probability measure, we can still apply the probabilistic argumentation framework, but we must take additional care when it comes to computing degrees of support and possibility. Examples of such under-specified probabilistic argumentation systems have been discussed in §3.3, where the question of Schema (1.1) has been interpreted in various ways from the perspective of the probabilistic argumentation framework. To deal with cases like that, let $\mathbb{P}$ denote an arbitrary set of probability measures over W, and let $\mathbb{A} = \{(V, \mathscr{L}_V, \Phi, W, P) : P \in \mathbb{P}\}$ be the corresponding family of probabilistic argumentation systems. For a given hypothesis ψ, we may then use each $\mathscr{A} \in \mathbb{A}$ to compute a pair of degrees of support and possibility, and then look at the sets $Y_{dsp} = \{dsp_{\mathscr{A}}(\psi) : \mathscr{A} \in \mathbb{A}\}$ and $Y_{dps} = \{dps_{\mathscr{A}}(\psi) : \mathscr{A} \in \mathbb{A}\}$ of all such degrees of support and possibility.

The immediate question that arises when applying this procedure, which comes out with a pair of target sets Y_{dsp} and Y_{dps}, is how to use those sets to judge the truth or falsity of the hypothesis ψ. By looking at degrees of support as a measure of cautiousness with respect to ψ being true in the light of the given evidence, we

R. Haenni et al., *Probabilistic Logics and Probabilistic Networks*, Synthese Library 350,   107
DOI 10.1007/978-94-007-0008-6_10, © Springer Science+Business Media B.V. 2011

suggest to select the most cautious value of Y_{dsp}, that is the *minimal* degree of support $dsp(\psi) = \min\{dsp_{\mathscr{A}}(\psi) : Y_{dsp}\}$. In order to be consistent with Definition 6 in §3.1, $\overline{dps}(\psi) = 1 - dsp(\neg\psi)$, we must then select the least cautious value of Y_{dps}, that is the *maximal* degree of possibility $\overline{dps}(\psi) = \min\{dps_{\mathscr{A}}(\psi) : Y_{dps}\}$. Note that $dsp(\psi) \leq \overline{dps}(\psi)$ still holds for all ψ, but since $dsp(\psi)$ and $\overline{dps}(\psi)$ will usually be generated by quite different elements of $\mathbb{P}$, we should refrain from taking them as respective bounds of a target interval that extends the idea of the standard semantics to probabilistic argumentation.

In the case where the set of probabilities $\mathbb{P}$ is restricted to a credal set, e.g. by employing a credal network for the specification of the admissible probability measures, it is guaranteed that both target sets Y_{dsp} and Y_{dps} are intervals. This follows from Definition 5, in which degrees of support are defined as conditional probabilities, and from the way convexity is transmittable to conditional probabilities (see §2.1.4 and §8.1.2).

10.2 Constructing and Applying the Credal Network

In §3 we discussed two different types of semantics for Schema (1.1). The semantics of the first type in §3.3.1 are all extensions of the standard semantics to the probabilistic argumentation framework, and we may thus follow a strategy similar to the one outlined in §9 to construct a network. But this is not the case for the semantics of the second type in §3.3.2, where each given probability set X_i is interpreted as a probability constraint for the reliability of a source providing the premise φ_i. Moreover, by assuming the sources to be pairwise independent, it seems that the second type of semantics does not provide adequate grounds for constructing non-trivial networks. The discussion in the remaining of this section will therefore not further pursue this particular way of interpreting instances of Schema (1.1).

In the extended standard semantics of §3.3.1, we assume a set Φ of logical constraints to be given in addition to the premises on the left hand side of Schema (1.1). For the premises to be useful to construct a network, let them include a set I of explicit probabilistic independencies, similar to §9.2. Note that such independencies may be available e.g. from representing a probabilistic argumentation system in form of Schema (1.1) as proposed in §3.2. In general, the best we can do then is to apply Algorithm 5 to the set of premises to get a credal network that represents the corresponding credal set w.r.t. Ω_W, where W denotes the set of variables appearing in the premises.

This credal network together with the common machinery from §8.2 can then be used to compute respective sets Y_{dsp} and Y_{dps} of degrees of support and possibility for a given hypothesis ψ. For this, the following algorithm consist of 3 steps: the first step is the network construction according to Algorithm 5, the second step is the determination of the sets $Args_{\mathscr{A}}(\psi)$, $Args_{\mathscr{A}}(\neg\psi)$, and $Args_{\mathscr{A}}(\bot)$, and the third step is the application of the common machinery from §8.2 to approximate the target intervals. Note that by explicitly calling the common machinery in last step, this algorithm differs from the algorithms of the following sections. The reason for this is that we need to adapt the hill-climbing algorithm to approximate degrees of support and possibility instead of probabilities of normal events.

Algorithm 6 Construction and evaluation of a credal network for the extended standard semantics, based on explicit independence assumptions and an additional set of logical constraints.

Input: Two sets $W = \{A_1, \ldots, A_M\}$ and $V \supseteq W$ of propositional variables, a set I of probabilistic independence relations (of the form $A_i \perp\!\!\!\perp A_j$ or $A_i \perp\!\!\!\perp A_j|A_k$) and premises $\varphi_1^{X_1}, \ldots, \varphi_N^{X_N}$ involving variables from W, a set Φ of logical constraints involving variables from V, and a query ψ.

1. **Network construction:** Use Algorithm 5 to construct a credal network that represents I and $\varphi_1^{X_1}, \ldots, \varphi_N^{X_N}$.
2. **Computing arguments, counter-arguments, and conflicts:** Use standard inference techniques to compute logical representations α^+, α^- and $\alpha^\perp$ of the sets $Args_{\mathscr{A}}(\psi), Args_{\mathscr{A}}(\neg\psi)$, and $Args_{\mathscr{A}}(\perp)$, respectively, where $\mathscr{A} = (V, \mathscr{L}_V, \Phi, W, P)$ is the involved probabilistic argumentation system (for some under-determined probability measure P).
3. **Target interval approximation:**

   a. Transform α^+, α^-, and α^- into disjoint sums-of-products (see §8.2.1).
   b. Compile the credal network (see §8.2.2) and instantiate it for each of the disjoint terms of α^+, α^-, and $\alpha^\perp$.
   c. Adapt and apply the hill-climbing algorithm from §8.2.3 to the compiled network to approximate lower and upper bounds of

$$dsp(\psi) = \frac{P(\alpha^+) - P(\alpha^\perp)}{1 - P(\alpha^\perp)} \quad \text{and} \quad dps(\psi) = \frac{1 - P(\alpha^-)}{1 - P(\alpha^\perp)}.$$

Output: Inner approximations of the target intervals Y_{dsp} and Y_{dps} for the first semantics of §3.3.1.

Note that one of the computationally hardest tasks of the above algorithm is the computation of the logical representations α^+, α^-, and $\alpha^\perp$ in the second step (their size is exponential in the worst case). Another difficult task is the transformation of α^+, α^-, and $\alpha^\perp$ into disjoint sums-of-products (there are exponentially many such terms in the worst case). We may thus need appropriate approximation algorithms, for example one that computes the shortest terms first (Haenni, 2002, 2003). By doing so, we may still get good approximations of the target intervals, but we must be aware that they may no longer be inner approximations.

With the above computational scheme, we obtain a way of extending the standard semantics to the probabilistic argumentation framework, but not for the further options given in §3.3.1. To extend it, for example for the semantics that takes each premise $\varphi_i^{X_i}$ as a constraint $dsp(\varphi) \in X_i$ on respective degrees of support, we must first compute respective logical representations α^i for each set $Args_{\mathscr{A}}(\varphi_i)$ and for

$Args_{\mathscr{A}}(\perp)$, and then construct the credal network with respect to the constraints $[P(\alpha^+) - P(\alpha^\perp)]/[1 - P(\alpha^\perp)] \in X_i$. To get rid of the non-linearity of those expressions, we may consider $k = 1 - P(\alpha^\perp)$ to be a constant value and then consider respective linear constraints $P(\alpha^+) - P(\alpha^\perp) \in kX_i$. This works in a similar way for constraints on degrees of possibility and for combined constraints (see §3.3.1).

Chapter 11
Networks for Evidential Probability

In §4 we discussed two different semantics for Schema (1.1) to answer two different types of inferential question for the theory of evidential probability. The first interpretation given to Schema (1.1) is just the usual semantics of evidential probability, which we call first-order EP (§4.3.1). The second interpretation given to Schema (1.1) is a proposal to evaluate the robustness of a particular first-order EP assignment, which we call second-order EP (§4.3.3).

Each semantics leads to very different properties of the entailment relation in Schema (1.1), however. First-order EP is a non-probabilistic, sub-System P non-monotonic logic of probabilities, whereas second-order EP is a genuine probabilistic logic. Thus the inferential methods for each semantics is quite different.

In this section we investigate how credal networks can be used to calculate an answer for second-order EP inference only, since first-order EP is essentially a logic of probability statements rather than a probabilistic logic. Nevertheless, second-order EP depends crucially upon outputs from first-order EP-arguments representing all possible counter-factual EP arguments with respect to the actually accepted evidence. So, we first present an algorithm for first-order EP, which does not rely upon credal networks, and then present an algorithm for constructing a credal network that may exploit the common machinery of §8.2.

11.1 First-Order Evidential Probability

Algorithm 7 Computing First-order Evidential Probability.

Input: a pair $\langle \chi, \Gamma_\delta \rangle$, where χ is a formula and $\Gamma_\delta = \{\varphi_1, \ldots, \varphi_n\}$ a set of formulas of $\mathscr{L}^{ep}$.

1. **Selection of potential probability statements:** Given $\langle \chi, \Gamma_\delta \rangle$, construct the set $\Gamma_{[\chi]}$ by Definition 7 in §4.1.1:

   a. Let $\Gamma_{[\chi]}$ be the equivalence class defined by

$$\{\%\mathbf{x}(\tau(\mathbf{x}),\rho(\mathbf{x}),[l,u]) : \chi \equiv \tau(\omega) \wedge \%(\tau(\omega),\rho(\omega),[l,u])\},$$

where each $\%\mathbf{x}(\tau(\mathbf{x}),\rho(\mathbf{x}),[l,u])$ of X is a direct inference statement of Γ_δ such that $\mathbf{x}$ is substituted by constants ω in τ and $\tau(\omega)$ is coextensive with χ.[1]

b. The direct inference statements in $\Gamma_{[\chi]} \subseteq \Gamma_\delta$ are the potential probability statements for χ.

2. **Selection of relevant statistical statements:** Given $\Gamma_{[\chi]}$, construct the set $\Gamma_{[\chi]}^{RS} \subseteq \Gamma_{[\chi]}$ by applying Richness and Specificity (§4.1.1):

a. Define $\Gamma_{[\chi]}^{R} \subseteq \Gamma_{[\chi]}$ such that $\forall \varphi, \vartheta \in \Gamma_{[\chi]}$, if φ and ϑ *conflict* and φ is a marginal distribution and ϑ a joint-distribution, then $\vartheta \in \Gamma_{[\chi]}^{R}$ but $\varphi \notin \Gamma_{[\chi]}^{R}$.

   i. **Determining conflict:** To determine whether direct inference statements φ, ϑ conflict, compare the interval $[l,u]$ of φ to the interval $[k,t]$ of ϑ. Only 3 comparisons are necessary. Given $[l,u]$: accept that $[k,t]$ conflicts with $[l,u]$ if:
   - $k < l$ and $t < u$, or
   - $l > k$ and $t > u$, or
   - $l = k$ and $t = u$.

   ii. **Distinguishing between a marginal distribution and a joint-distribution:** There is a *syntactic* difference between a marginal distribution whose statistic is n dimensions and a joint-distribution whose statistic is $n + m$ dimensions, for some positive m. This will be expressed by the arity of the target formula predicates (τ) appearing in φ and ϑ.

b. Define $\Gamma_{[\chi]}^{RS} \subseteq \Gamma_{[\chi]}^{R}$ such that $\forall \varphi, \vartheta \in \Gamma_{[\chi]}^{R}$, if φ and ϑ *conflict* but ρ of φ and ρ' of ϑ are such that $\rho \subset \rho'$, then $\varphi \in \Gamma_{[\chi]}^{RS}$ but $\vartheta \notin \Gamma_{[\chi]}^{RS}$.

   i. Apply step 2.a.i. to determine conflict, replacing $\Gamma_{[\chi]}$ with $\Gamma_{[\chi]}^{R}$.

   ii. **Selecting specific statistics for χ:** Specificity concerns explicit logical relationships between reference classes ρ of φ and ρ' of ϑ that appear in Γ_δ. Specificity applies to φ and ϑ when both are in $\Gamma_{[\chi]}^{R}$, the interval of φ conflicts with the interval of ϑ but $\rho \subset \rho'$.

3. **Find the shortest cover:** Given $|\Gamma_{[\chi]}^{RS}| = n$, define $\langle X_1, X_2, \ldots, X_n \rangle$, where l_i and u_i are the lower bound and the upper bound, respectively, of the ith interval X_i in $\mathscr{I}$ as the set of n intervals expressed by the formulas in $\Gamma_{[\chi]}^{RS}$. Apply Strength (§4.1.1) to $\mathscr{Y}$ closed under difference by first defining:

(i) $L = \langle X_1', X_2', \ldots, X_n' \rangle$ to be a permutation of $\mathscr{I}$ such that for all $j > i$, either $l_i' < l_j'$ or $\left(l_i' = l_j' \text{ and } u_i' \geq u_j' \right)$.

(ii) $U = \langle X_1'', X_2'', \ldots, X_n'' \rangle$ to be a permutation of $\mathscr{I}$ such that for all $j > i$, either $u_i'' > u_j''$ or $\left(u_i'' = u_j'' \text{ and } l_i'' \leq l_j'' \right)$.

---

[1] Note that the sequence of constants ω must be such that the variable substitution of $\tau(\omega)$ leaves no free variables in τ, but that $\rho(\omega)$ may have open variables. This will be important in the implementation of the Richness rule. Also, we relax notation here by omitting the corner-quotes.

Then:

a. If $|\mathscr{I}| = 1$, the closure under conflict of $\mathscr{I} = \mathscr{I}$, i.e., $CC(\mathscr{I}) = \{X_1\}$.
b. Otherwise, define $S_0 = \{X'_n, X''_n\}$ and the smallest cover of S_0 as $[l_0, u_0]$. Let $i = 0$.
   i. Then define S_{i+1} as the set $S_i \cup \{X_k \in \mathscr{I}$: either the lower bound of X_k is greater than the lower bound of S_i, $l_k > l_i$, or the upper bound of X_k is less than the upper bound of S_i, $u_k < u_i\}$.
   ii. Repeat until $S_{i+1} = S_i$, in which case denote by S.
   iii. Then, $CC(\mathscr{I}) = S$.

Output: An interval $[l, u]$ corresponding to the proportion of EP models of $\varphi_1^1, \ldots, \varphi_n^1$ that are also models of χ.

11.2 Second-Order Evidential Probability

The first-order question concerns the assignment of evidential probability to a statement given some evidence, which is captured in Schema (1.1) by the EP semantics for $\models$. The second-order question concerns the proportion of overlap among all possible combinations of evidence for ψ occurring in Γ_δ.

As discussed in §4.3.3, the second-order EP probability of ψ on possibly relevant evidence $\varphi_1, \ldots, \varphi_m$ is

$$P(\psi) = \sum_{e_1, \ldots, e_m = 0}^{1} P\left(\psi | \varphi_1^{e_1}, \ldots, \varphi_m^{e_m}\right) \cdot P\left(\varphi_1^{e_1}, \ldots \varphi_m^{e_m}\right),$$

i.e., summing over all possible evidence for ψ with respect to Γ_δ, where $P(\varphi_i) \in [1 - \delta, 1]$ is the prior confidence in the evidence for ψ. We assume in this book that items of evidence are independent in the absence of evidence of dependence, in which case

$$P\left(\varphi_1^{e_1}, \ldots \varphi_m^{e_m}\right) = \prod_{1}^{m} P\left(\varphi_i^{e_i}\right),$$

where $P(\varphi_i) \in [\lambda, 1]$, $\lambda = 1 - \delta$.[2]

A graphical representation of the dependency relations in a second-order EP inference is represented by the edges in Figure 11.1. ψ, which is a statement about a first-order evidential probability, can only depend on the set $\{\varphi_1, \ldots, \varphi_m\}$ of possibly relevant evidence (as defined in §4.3.2). Hence the possibly relevant evidence statements screen off ψ from other evidential statements $\varphi_{m+1}, \ldots, \varphi_n$, and can be taken as the parents of ψ in a credal network. Under the assumption that these evidence

---

[2] Note that details about the theory of acceptance are relevant in selecting λ. If we treat the interval $[1 - \delta, 1]$ as characterizing the risk of accepting the *set* of premises, assign equal probability of risk of error to each premise while assuming these probabilities are independent, then $\lambda = 1 - \delta^{\frac{1}{m}}$. But the risk term δ is more usually thought to characterize the risk of accepting *each individual* statement; this is interpretation adopted here.

statements are independent, no further arrows are required to link the evidential statements.

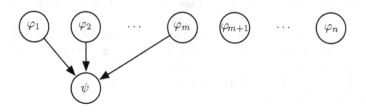

Fig. 11.1: Second-order EP credal network.

We now present an algorithm for calculating second-order EP.

Algorithm 8 Computing Second-order Evidential Probability.

Input: $\varphi_1^{X_1},\ldots,\varphi_n^{X_n} \models \psi$?, where ψ is $P(\chi) \in [l,u]$, construed as the first-order EP assignment $\langle \chi,[l,u] \rangle$, on the premises $\Gamma_\delta = \varphi_1,\ldots,\varphi_n$, that is output by Algorithm 7.

1. **Construction of graph:**

    a. Construct the set $\{\varphi_1,\ldots,\varphi_m\} = \Delta \subseteq \Gamma_\delta$ of possibly relevant evidence for χ with respect to Γ_δ (§4.3.2).

    b. Define the set of nodes V to include $\varphi_1,\ldots,\varphi_n,\psi$.

    c. Connect arrows from each $\varphi_1,\ldots,\varphi_m$ to ψ. (See Figure 11.1).

2. **Probabilities:**

    a. Attach: X_1 to $\varphi_1,\ldots,X_n$ to φ_n.

    b. Compute conditional probabilities: $P\left(\chi | \varphi_1^{e_1},\ldots,\varphi_m^{e_m}\right) = \frac{\|[l,u] \cap [l',u']\|}{\|[l,u]\|}$, when $e_i \in \{0,1\}$ for all $1 \leq i \leq m$.

Output: a credal network with graph $\mathcal{G}$, representing probabilistic independence relations implicit in second-order EP inference, and probabilities calculated as described above.

To compute Y for ψ, use the common machinery of §8.2.

For example, in second-order evidential probability we might be faced with the following question

$$\%x(Fx,Rx,[.2,.4])^{[.9,1]}, Rt \approx P(Ft) \in [.2,.4]?$$

This asks, given evidence that (i) the proposition that the frequency of attribute F in reference class R is between .2 and .4 has probability at least .9, and (ii) t falls in reference class R, what probability interval should attach to the proposition that the probability that t has attribute F is between .2 and .4? In first-order EP, if $1 - \delta \geq .9$ then $\text{Prob}(Ft, \Gamma_\delta) = [.2, .4]$ would be conclusively inferred (and hence treated as if it had probability 1). Clearly this disregards the uncertainty that attaches to the statistical evidence; the question is, what uncertainty should attach to the conclusion as a consequence? (This is a second-order uncertainty; hence the name *second-order* evidential probability.) One can construct a credal network to answer this question as follows. Let φ_1 be the proposition $\%x(Fx, Rx, [.2, .4])$, φ_2 be Rt and ψ be $P(Ft) \in [.2, .4]$. These can all be thought of as variables that take possible values True and False. The structure of 1oEP calculations determines the structure of the directed acyclic graph in the credal network:

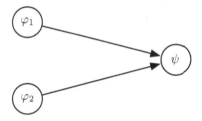

Fig. 11.2: Credal net for 2oEP.

(Note that in this example all evidence is possibly relevant evidence, $m = n$.) The conditional probability constraints involving the premiss propositions are simply their given risk levels:

$$P(\varphi_1) \in [.9, 1],$$
$$P(\varphi_2) = 1.$$

Turning to the conditional probability constraints involving the conclusion proposition, these are determined by our assumption that first-order evidential probability is uniformly distributed in the evidential probability interval (§4.3.3):

$$P(\psi | \varphi_1 \wedge \varphi_2) = 1,$$

$$P(\psi | \neg \varphi_1 \wedge \varphi_2) = P(\psi | \varphi_1 \wedge \neg \varphi_2) = P(\psi | \neg \varphi_1 \wedge \neg \varphi_2) = .2.$$

Finally, the Markov condition holds in virtue of our default assumption of independent evidence, which implies that $\varphi_1 \perp\!\!\!\perp \varphi_2$. Inference algorithms for credal networks can then be used to infer the uncertainty that should attach to the conclusion, $P(\psi) \in [.92, 1]$. Hence we have:

$$\%x(Fx, Rx, [.2, .4])^{[.9,1]}, Rt \approx P(Ft) \in [.2, .4]^{[.92,1]}$$

11.3 Chaining Inferences

While it is not possible to chain inferences in first-order EP, this is possible in second-order EP, and the credal network representation can just as readily be applied to this more complex case. Consider the following question:

$$\%x(Fx, Rx, [.2, .4])^{[.9,1]}, Rt, \%x(Gx, Fx, [.2, .4])^{[.6,.7]} \approx P(Gt) \in [0, .25]?$$

As we have just seen, the first two premisses can be used to infer something about Ft, namely $P(Ft) \in [.2, .4]^{[.92,1]}$. But now this inference can then be used in conjunction with the third premiss to infer something about Gt. To work out the probability bounds that should attach to an inference to $P(Gt) \in [0, .25]$, we can apply the credal network procedure. Again, the structure of the graph in the network is given by the structure of EP inferences:

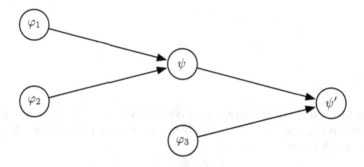

Fig. 11.3: Credal net for chaining inferences in 2oEP.

Here φ_3 is $freq_F(G) \in [.6, .7]$ and ψ' is $P(Gt) \in [0, .25]$; other variables are as before. The conditional probability bounds of the previous example simply carry over

$$P(\varphi_1) \in [.9, 1], P(\varphi_2) = 1,$$

$$P(\psi|\varphi_1 \wedge \varphi_2) = 1, P(\psi|\neg\varphi_1 \wedge \varphi_2) = .2 = P(\psi|\varphi_1 \wedge \neg\varphi_2) = P(\psi|\neg\varphi_1 \wedge \neg\varphi_2).$$

But we need to provide further bounds. As before, the risk level associated with the third premiss φ_3 provides one of these:

$$P(\varphi_3) \in [.6, .7],$$

and the constraints involving the new conclusion ψ' are generated as follows:

$$P(\psi'|\psi \wedge \varphi_3) = \frac{|[.2 \times .6 + .8 \times .1, .4 \times .7 + .6 \times .1] \cap [0, .25]|}{|[.2 \times .6 + .8 \times .1, .4 \times .7 + .6 \times .1]|} = .31,$$

$$P(\psi'|\neg\psi \wedge \varphi_3) = .27, P(\psi'|\psi \wedge \neg\varphi_3) = P(\psi'|\neg\psi \wedge \neg\varphi_3) = .25.$$

The Markov Condition holds in virtue of evidential independence and the structure of EP inferences. Performing inference in the credal network yields $P(\psi') \in [.28, .29]$. Hence

$$\%x(Fx, Rx, [.2, .4])^{[.9,1]}, Rt, \%x(Gx, Fx, [.2, .4])^{[.6,.7]} \approx P(Gt) \in [0, .25]^{[.28,.29]}.$$

This example shows how general inference in 2oEP can be: we are not asking which probability bounds attach to a 1oEP inference in this example, but rather which probability bounds attach to an inference that cannot be drawn by 1oEP. The example also shows that the probability interval attaching to the conclusion can be narrower than intervals attaching to the premisses.

Chapter 12
Networks for Statistical Inference

In §5 we discussed two ways in which classical statistics can be captured in Schema (1.1), one using functional models and fiducial probability, and one using evidential probability to represent the fiducial argument. This section investigates the use of the common machinery of §8.2 and Algorithm 7 in classical statistics.

12.1 Functional Models and Networks

As described in §5.1.2, one way of capturing classical statistics in Schema (1.1) makes use of fiducial probability and functional models: on the basis of an invertible functional relation $f(H_\theta, \omega) = d$ between data D, stochastic elements ω, and hypotheses H_θ, a probability assignment $P(\omega)$ determines a probability assignment $P(H_\theta)$. While the use of fiducial probability in Schema (1.1) is rather limited, we want to present a way in which credal networks may be used to aid the computation of fiducial probability.

12.1.1 Capturing the Fiducial Argument in a Network

To set the stage, we first provide a network representation for the fiducial inference represented by functional models. We restrict attention to fiducial probability using an invertible functional relation. It may be recalled from §5.1.2 that we can relax the requirement of invertibility for the functional relation f, and derive degrees of support and possibility for a hypothesis H_I consisting of a set of hypotheses H_θ. However, we will not discuss the generalisation to degrees of support and possibility here.

In a functional model representing the fiducial argument, the relation between the hypotheses H_θ, the stochastic elements ω, and the data D is such that the stochastic elements and the hypotheses are probabilistically independent:

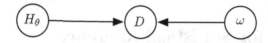

Fig. 12.1: A network representing the conditional independencies between H_θ, ω and D in the functional model $f(H_\theta, \omega) = D$ that generates the fiducial probability for H_θ from the distribution over the stochastic elements ω.

$$P(H_\theta, \omega) = P(H_\theta)P(\omega). \tag{12.1}$$

Moreover, given a hypothesis h_θ and a stochastic element ω, the occurrence of the data d is completely determined, $P(d|h_\theta, \omega) = I_d(h_\theta, \omega)$ where $I_d(h_\theta, \omega) = 1$ if $f(h_\theta, \omega) = d$ and $I_d(h_\theta, \omega) = 0$ otherwise. The corresponding network is depicted in Figure 12.1.

If we condition on the observed data d then, because of the network structure and the further fact that the relation $f(H_\theta, \omega)$ is deterministic, the variables ω and H_θ become perfectly correlated: each h_θ is associated with some $\omega = f^{-1}(h_\theta, d)$. Assuming that the observation of d does not itself influence the probability of ω, meaning that $P(\omega|d) = P(\omega)$ we can therefore write

$$P(H_\theta|d) = P(f^{-1}(H_\theta, d)) = P(\omega). \tag{12.2}$$

This means that after observing d we can transfer the probability distribution over ω to H_θ according to the function f^{-1}. In other words, the crucial step in the fiducial argument is here seen as a special case of the fact that conditioning on a collision node creates a correlation between the parent nodes.

12.1.2 Aiding Fiducial Inference with Networks

We now turn to the use of networks in aiding fiducial inference. The natural suggestion is to consider cases in which there are many different statistical parameters θ_j in the model, and equally many observed variables D_j that may be used to perform fiducial arguments. We may then employ independence relations that obtain between the parameters and the data to speed up the fiducial inference.

Example 20. Consider hypotheses H_θ determined by two variables, $\theta = \langle \theta_1, \theta_2 \rangle$, stochastic elements $\omega = \langle \omega_1, \omega_2 \rangle$, and two propositions D_1 and D_2 representing the data. Say that we have a number of independence relations between these variables and data sets:

$$P(D_1 \wedge D_2|H_\theta \wedge \omega) = P(D_1|H_{\theta_1} \wedge \omega_1)P(D_2|H_{\theta_1} \wedge H_{\theta_2} \wedge \omega_2), \tag{12.3}$$

$$P(H_\theta) = P(H_{\theta_1})P(H_{\theta_2}), \tag{12.4}$$

$$P(\omega) = P(\omega_1)P(\omega_2). \tag{12.5}$$

Finally, we have smoothly invertible functional relations $f(H_{\theta_1}, \omega_1) = D_1$ and $f_{\theta_1}(H_{\theta_2}, \omega_2) = D_2$, meaning that for each fixed value of θ_1 the function f_{θ_1} is smoothly invertible. The corresponding network is depicted in Figure 12.2. We first

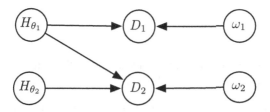

Fig. 12.2: A network representing the conditional independencies between H_θ, ω and D as expressed in Equations (12.3) to (12.5).

derive a fiducial probability $P(H_{\theta_1}|d_1)$, whereby we assume that we do not yet have knowledge of D_2 or H_{θ_2}. After that we derive a fiducial probability for H_{θ_2} from d_2, by first computing a fiducial probability over H_{θ_2} for each value of θ_1 separately. We then use the law of total probability to arrive at $P(H_{\theta_2}|d_2)$.

This is a version of the so-called step-by-step method for the fiducial argument. See Seidenfeld (1992). The method was devised by Fisher for deriving the fiducial probability over statistical hypotheses characterised by more than one parameter. We can easily generalise the application of networks to step-by-step fiducial inference for larger numbers of statistical parameters, data sets, and stochastic elements.

Algorithm 9 Construction of a network for fiducial inference over more than one statistical parameter.

Input: The following assumptions to run the step-by-step fiducial argument.

1.   A sample space Ω_D involving propositional variables $D = \{D_1, \ldots, D_N\}$, a model Ω_H of hypotheses H_θ with $\theta = \langle \theta_1, \ldots, \theta_N \rangle$, and a space of stochastic elements $\Omega_W = \langle \omega_1, \ldots \omega_N \rangle$.

2.   Smoothly invertible functional relations $f_j^{S_j}(H_{\theta_j}, \omega_j) = D_j$ in which $S_j \subseteq \{\theta_{j'} : j' < j\}$, such that step-by-step fiducial inference can be supported. The sets S_j determine which hypotheses need to be fixed for the functional relations $f_j^{S_j}$ to be smoothly invertible.

3.   A set of independence relations I among the marginal probabilities $P(H_{\theta_j})$ of the hypotheses. Next to $P(D_1|H_\theta) = P(D_1|H_{\theta_1})$, this set must include the following independence relations for each $j > 1$:

$$P\left(D_j \wedge D_{j-1} | H_\theta\right) = P\left(D_{j-1} \Big| \bigwedge_{j' \in S_{j-1}} H_{\theta_{j'}}\right) P\left(D_j \Big| \bigwedge_{j' \in S_j} H_{\theta_{j'}}\right),$$

$$P(H_\theta) = \prod_j P(H_{\theta_j}),$$

$$P(\omega) = \prod_j P(\omega_j),$$

Construction of the network: run Algorithm 5 with the independence relations I to arrive at a network for fiducial inference.

Output: a network structure in which fiducial inference can be performed.

It must be conceded that the algorithm for setting up a network for fiducial inference requires a lot of input. This is because the fiducial argument itself can only be run under fairly specific circumstances. Still, once the network is built, it will structure and possibly simplify the calculation of fiducial probability. The idea is again that once the network has been built, computation is much less costly.

12.1.3 Trouble with Step-by-Step Fiducial Probability

As already indicated at the start of this section, fiducial probabilities are controversial, and this is all the more so for the step-by-step procedure. We explain briefly why we recommend caution in the application of the step-by-step fiducial argument.

Returning to the example, note that we can only apply the fiducial argument on the condition that, in the absence of knowledge about H_{θ_2}, the data D_1 indeed allow for transferring the probability assignment over ω_1 to the hypotheses H_{θ_1}. Now it seems that this possibility is guaranteed by the use of the smoothly invertible functional relation $f(H_{\theta_1}, \omega_1) = D_1$. But there is something a bit awkward about the assumption of the absence of knowledge about H_{θ_2} and the use of the function f, which is brought out in Seidenfeld (1979).

Imagine that we do have some fixed probability assignment $P(H_{\theta_2})$ before we apply the fiducial argument to derive $P(H_{\theta_1})$ from d_1. Now consider the graph in Figure 12.2. From this graph we can deduce that if there is some definite probability assignment $P(H_{\theta_2})$, learning about D_2 will be probabilistically relevant to the probability of H_{θ_1}. Of course, there are special cases of probability assignments $P(H_{\theta_2})$ for which the data on D_2 come out as irrelevant anyway, but this is not the case in general. Hence, by the mere assumption of a particular distribution $P(H_{\theta_2})$ we might destroy the smoothly invertible relation $f(H_{\theta_1}, \omega_1) = D_1$ and replace it with $f(H_{\theta_1}, \omega) = D_{12}$, where D_{12} refers to the combination of D_1 and D_2: we need both these data to derive a probability assignment over H_{θ_1}. In this case the fiducial argument cannot be run, so a lot hinges on the absence of knowledge about H_{θ_2}.

This makes step-by-step fiducial inference more problematic than its simple and direct application. Moreover, there is no indication that network structures can help to solve the problem, although they may help getting clear on framing the problem itself. That is, the network of §12.1.1 might provide a helpful perspective on the fiducial argument, but this perspective does not lead naturally to a solution for the difficulties associated with the step-by-step procedure. Whether a more elaborate representation of the functional relations in terms of networks would do better on this count is a question for further research.

12.2 Evidential Probability and the Fiducial Argument

The novelty of the fiducial argument is to treat indirect inference as a form of direct inference, and the approach that EP takes is to use knowledge about the sample and the sampling procedure to eliminate those reference statistics from which we have reason to believe are not representative. Since there is no network structure for first-order EP, the role that absence of knowledge plays in the fiducial argument is viewed not as an awkward feature of Fisher's argument but rather a central feature of statistical inference as such. We noted in §5.1.3 the clash between EP's logic of probabilities and probabilistic logic, and with it the prospects of viewing classical statistical inference in terms of a probabilistic logic. But the view does give a picture of statistical inference as a form of uncertain inference, and we may use the machinery developed in §11 to calculate fiducial probabilities.

12.2.1 First-Order EP and the Fiducial Argument

To interpret
$$\varphi_1^{X_1}, \ldots, \varphi_n^{X_n} \mathrel{|\!\approx} \psi^?,$$
as an EP reconstruction of fiducial inference, we replace $\mathrel{|\!\approx}$ by $\mathrel{|\!\sim}$ of first-order EP-entailment and interpret the premises to include logical information and direct inference statements, just as we do §11, and add a selection function $f(\chi, U(x)_{[x/\omega]}) \geq \lambda$ that restricts the set of relevant direct inference statements to just those that are rationally representative of χ to at least degree λ.

As we observed in §11, first-order EP does not admit the use of the common machinery of §8.2. We instead have Algorithm 7 to calculate first-order EP assignments. Thus there is no inferential machinery called upon apart from what is provided by first-order EP.

As we observed in §5.3, this approach does not resolve the controversy surrounding the fiducial argument but instead pinpoints the where the controversy lies, which is how to restrict (or, rather, rule out) otherwise applicable reference statistics. Viewed in terms of a first-order EP inference, this restriction will involve logical formulas in our knowledge base that fix a subset of relevant statistics to accept as rationally representative for a particular assignment $[l, u]$ to χ.

12.2.2 Second-Order EP and the Fiducial Argument

The first-order question concerns the assignment of evidential probability to a statement under the restriction of the pair (f, λ), where f is the 'rationally-representative to degree λ' selection function $f(\chi, U(x)_{[x/\omega]}) \geq \lambda$ on the relevant statistics for a sentence χ. The second-order question concerns the proportion of overlap among all possible combinations of evidence for ψ occurring in Γ_δ under the restriction (f, λ).

The second-order EP probability of ψ restricted by (f, λ) on the knowledge base $\Gamma_\delta = \left\{ \varphi_1^{X_1}, \ldots, \varphi_n^{X_n} \right\}$ is

$$P(\psi) = \sum_{e_1, \ldots, e_m = 0}^{1} P\left(\psi | \varphi_1^{e_1}, \ldots, \varphi_m^{e_m}\right) \cdot P\left(\varphi_1^{e_1}, \ldots \varphi_m^{e_m}\right),$$

where $\sum_{e_1, \ldots, e_m = 0}^{1} P\left(\psi | \varphi_1^{e_1}, \ldots, \varphi_m^{e_m}\right)$ sums over all possible evidence given the restriction imposed by (f, λ) for ψ with respect to Γ_δ. We replace $P\left(\varphi_1^{e_1}, \ldots \varphi_m^{e_m}\right)$ by

$$\prod_1^m P\left(\varphi_i^{e_i}\right),$$

where $P(\varphi_i) \in [\lambda, 1]$, $\lambda = 1 - \delta$.

Once (f, λ) picks $\left\{ \varphi_1^{X_1}, \ldots, \varphi_m^{X_m} \right\}$ from $\left\{ \varphi_1^{X_1}, \ldots, \varphi_n^{X_n} \right\}$, the algorithm for calculating the second-order EP probability for a fiducial argument is just the algorithm for second-order EP, Algorithm 8.

Chapter 13
Networks for Bayesian Statistical Inference

The essential property of Bayesian statistical inference, as introduced in §6, is that the probability space includes both the sample space Ω_D, consisting of propositional variables D_i, and the space of hypotheses Ω_H. In the semantics of Bayesian statistical inference, the premises include a prior over hypotheses $P(H_j)$ and the likelihoods of the hypotheses $P(D_i|H_j)$. The conclusion is a posterior over hypotheses $P(H_j|d_s^e)$ for some observed assignment d_s^e to the D_i, where e is a vector of assignments and s is a corresponding vector of values for i. In the following we will present two ways in which credal networks may be employed in this inference. The first presents a computational improvement of the inferences as such, while the second supplements the inference with further tools, deriving from the common machinery of §8.

13.1 Credal Networks as Statistical Hypotheses

We first spell out how a credal network can be related to a statistical model, i.e. a set of statistical hypotheses. Recall that a credal network is associated with a credal set, a set of probability functions over some designated set of variables. Hence a credal set may be viewed as a statistical model: each element of the credal set is a probability function over the set of variables, and this probability may be read as a likelihood of some hypothesis for observations of valuations of the network. Conversely, any statistical model concerns inter-related trials of some specific set of variables, so that we can identify any statistical model with a credal network containing these variables. Here we deal with non-causal statistical hypotheses; (Leuridan, 2008, Chapter 4) argues that credal nets can also be used to represent causal hypotheses. An detailed illustration of many ideas in this section can be found in (Romeijn et al., 2009).

R. Haenni et al., *Probabilistic Logics and Probabilistic Networks*, Synthese Library 350, DOI 10.1007/978-94-007-0008-6_13, © Springer Science+Business Media B.V. 2011

13.1.1 Construction of the Credal Network

To construct the credal network that represents a statistical model, we first list all the independence relations that obtain between the variables in the sample space Ω_D.

Example 21. Consider subsequent observations, at times i, of three binary variables $V_i = \{A_i, B_i, C_i\}$. The independence relations I are as follows:

$$\forall i : A_i \perp\!\!\!\perp B_i | C_{i-1},$$
$$\forall i' \neq i-1 : A_i \perp\!\!\!\perp C_{i'}, B_i \perp\!\!\!\perp C_{i'},$$
$$\forall i' \neq i : C_i \perp\!\!\!\perp C_{i'}, A_i \perp\!\!\!\perp A_{i'}, B_i \perp\!\!\!\perp B_{i'}, A_i \perp\!\!\!\perp B_{i'}, B_i \perp\!\!\!\perp B_{i'}.$$

To determine the credal network, we run the algorithm of section §9.2, which also allows us to include further constraints $\varphi_i^{X_i}$ on the probability assignments of the variables.

Algorithm 10 Construction of a credal network for the semantics of Bayesian statistical inference, based on a given model of statistical hypotheses.

Input: a model Ω_H of statistical hypotheses H_θ, a sample space Ω_D concerning a set $V = \{A_1, \ldots, A_M\}$ of propositional variables, and further premises $\varphi_1^{X_1}, \ldots, \varphi_N^{X_N}$ involving those variables.

1. **Derivation of the independence relations:** using the (conditional) independence relations inherent to the likelihoods of all the H_θ

   a. If $P(A_i \wedge A_j \mid H_\theta) = P(A_i \mid H_\theta)P(A_j \mid H_\theta)$, add $A_i \perp\!\!\!\perp A_j$ to the set of independence constraints I.

   b. If $P(A_i \wedge A_j \mid \bigwedge_k A_k \wedge H_\theta) = P(A_j \mid \bigwedge_k A_k \wedge H_\theta)P(A_j \mid \bigwedge_k A_k \wedge H_\theta)$, add $A_i \perp\!\!\!\perp A_j \mid \bigwedge_k A_k$ to the set of independence constraints I.

2. **Running Algorithm 5:** based on the set I of independence relations and the premises $\varphi_1^{X_1}, \ldots, \varphi_N^{X_N}$.

Output: a graph $\mathcal{G}$ and a set of extremal points v_i in terms of the network coordinates.

As the corresponding credal set we take the complete extension of the network. Under this extension, the credal set is basically a set of Bayesian networks that all have the given credal network as their graph. The fact that the statistical model is thus associated with a Bayesian network containing free coordinates will turn out to be useful in learning from data.

Now that we have identified credal sets as statistical models, the notion of a second-order probability assignment falls into place quite naturally. The second-order probability assignment takes the members of the credal set as arguments. Any credal set may be captured by a second-order probability over probability functions of the variables that is non-zero only at those functions belonging to the credal set. A uniform distribution is perhaps the most natural choice, but other shapes of the second-order probability are possible too.

Example 22 (continued). Recall example Example 21. Even if we restrict attention to Markov processes with these variables, a statistical hypothesis on them is determined by $2^3(2^3 - 1) = 56$ probabilities for the transitions between valuations of the variables. So the corresponding statistical model Θ has a coordinate system with 56 coordinates. Now consider the network depicted in Figure 13.1, which captures the independence relations.

Fig. 13.1: The graph capturing the independence relations among the propositional variables A_i, B_i, and C_i. This graph holds for each and every value of i.

The complete extension of this network consists of the probability functions arrived at by filling in the probabilities on the nodes and edges. Again each of these functions is associated with a statistical hypothesis, denoted H_η. The coordinate system of the corresponding model Θ_{net} is a vector $\eta = \langle \gamma, \alpha_0, \beta_0, \alpha_1, \beta_1 \rangle$, where

$$P(c_i | H_\eta) = \gamma,$$

$$P\left(a_{i+1} | c_i^k \wedge H_\eta\right) = \alpha_k,$$

$$P\left(b_{i+1} | c_i^k \wedge H_\eta\right) = \beta_k.$$

The space Θ_{net} is a strict subset of the space Θ. The above credal network may be characterised by a prior probability assignment that is nonzero only on this subset.

Against the background of a statistical model as the complete extension of a credal network, we first discuss Bayesian statistical inference as it has been presented in the foregoing. For a more complete example of the application of credal networks in the context of Bayesian inference over both latent and observable variables, see (Romeijn et al., 2009). Section §13.2 concerns an extension of Bayesian statistical inferences on this basis.

13.1.2 Computational Advantages of Using the Credal Network

We consider inferences to do with adapting the second-order probability assignment over the credal set according to Bayesian statistical inference, on the basis of repeated observations of the variables.

One important consequence of viewing the statistical model as a credal network is that it allows us to reduce the dimensions of the statistical model, as the above example illustrates: the credal network restricts the space of allowed parameter values from $\dim(\Theta) = 56$ to $\dim(\Theta_{net}) = 5$ dimensions. This reduction in the dimensions of the model can entail major reductions in computational load, parallel to the reduction in computational load effected by using Bayesian networks when dealing with single probability assignments.

Another important consequence concerns the coordinate system of the statistical model. Note that by replacing the product of simplexes Θ with the space Θ_{net}, we are not only reducing the number of coordinates, but we are also availing ourselves of coordinates that are orthogonal to each other. If we want to update the probability assignment over the model with the observation of one variable, while the observations of other variables are unknown, this can more easily be captured analytically in the coordinate system deriving from the credal network. In particular, if we start out with a prior probability density over the model that is factorisable, $P(H_\eta) = f_C(\gamma) f_{A_0}(\alpha_0) f_{B_0}(\beta_0) f_{A_1}(\alpha_1) f_{B_1}(\beta_1)$, updates on the variables run entirely separately. So if we find the assignment c_i after some sequence of particular observations d, we need only adapt the density factor $f_C(\gamma|d)$ to $f_C(\gamma|d \wedge c_i)$. The other factors are invariant under this update.

This computational advantage shows up in the predictive probabilities that derive from the posterior probability assignments over the model. As indicated in §6, if the probability density over statistical hypotheses is a Dirichlet distribution, then the predictions for valuations of the separate network variables are Carnapian prediction rules. Consequently, entire network valuations are predicted according to products of such prediction rules, as is worked out in more detail in Romeijn (2006). In there it is further explained that the use of the coordinate system deriving from the network helps us to define and motivate alternative prior probability assignments over the model, allowing us to capture variations in inductive influences among valuations of the network. In Romeijn (2005) these insights are generalised to prior probability densities outside the family of Dirichlet priors.

In sum, credal networks can be used to improve computation in Bayesian statistical inferences. However, there is not a direct application of the common inference machinery of §8 to this use of credal networks. Rather the present usage remains internal to the inference machinery provided by Bayesian statistics itself, which employs probability density functions over the credal set. In the remainder of this section we turn to an application of credal networks that directly employs the common inference machinery. This involves a genuine extension of Bayesian statistical inference itself.

13.2 Extending Statistical Inference with Credal Networks

In this section we include the statistical hypotheses in a credal network representation, and investigate some of the evidential and logical relations between hypotheses and observed variables.

13.2.1 Interval-Valued Likelihoods

In the foregoing we have associated statistical hypotheses with probability assignments in the credal set: every hypothesis is associated with a single such likelihood function, and thus with a single probability assignment over the observable variables. But we may also add the hypotheses as nodes in the network.

Example 23. Consider a credal network consisting of a hypothesis node H_j with $j \in \{0,1\}$ and a large number of instantiation nodes of the propositional variables C_i, labelled with i. This credal network is depicted in Figure 13.2.

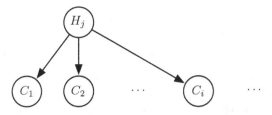

Fig. 13.2: A graph containing the statistical hypothesis H_j as the top node. Conditional on the hypothesis, the propositional variables C_i for $i = 1, 2, \ldots$ are independent.

These instantiation variables are independent of each other conditional on H_j, and each value j may be associated with a conditional probability of each of the instantiations. These conditional probability assignments can be filled in on the edges leading from the hypothesis node H_j to the separate instantiations of C_i.

The idea to include a hypothesis in a credal network leads quite naturally to the idea of interval-valued likelihoods. That is, we may assign a probability interval to the edges connecting a statistical hypothesis with the data. Illustrating this by the above example, we replace the sharp probability values for c_i with

$$P(c_i|H_0) \in [0.3, 0.7], \tag{13.1}$$
$$P(c_i|H_1) \in [0.6, 0.8]. \tag{13.2}$$

In words, this expresses that the statistical hypotheses are not exactly clear on the probability of c_i, although they do differ on it. The use of this formal possibility is in a sense complementary to the well-known method of Jeffrey conditioning. Jeffrey conditioning tells us what to do in case we are not quite sure what the evidence we gathered is, while we are sure how the evidence, once made precise, bears on the hypotheses; this is expressed in the condition of rigidity. The interval-valued likelihoods, by contrast, provide us with a technique to update probability assignments in case we know exactly what evidence we have gathered, while we are not sure how this evidence bears on the hypotheses at issue.

The idea of interval-valued likelihoods is not new; it has already been dealt with by Halpern and Pucella (2005). There the updates of probability assignments are carried out using Dempster's rule of combination. The advantage of presenting the idea in the context of Schema (1.1) is that we can directly use the common inference machinery. Starting with an unknown probability for both H_j, for example $P(H_j) \in [0,1]$, we may input the instantiations of several c_i, to arrive at new interval-valued probability assignments for both hypotheses. The common machinery of credal networks, as worked out in §8, can thus be applied directly to the statistical setting.

Algorithm 11 Construction of a credal network for Bayesian statistical inference in which the statistical hypotheses are included.

Input: a finite set of credal sets $\mathbb{K}_j$ for $j = 1, \ldots N$, each based on a given set of propositional variables $V = \{A_1, \ldots A_M\}$, independence relations I, and premises $\varphi_i^{X_i}$, such that the sets are all associated with a unique credal network $\mathscr{G}$.

1. **Addition of hypotheses nodes:** add the node H_j with possible valuations $j = 1, \ldots N$, and connect the node H_j with each variable A_j in the graph $\mathscr{G}$.
2. **Fixing the likelihoods:** associate each credal set $\mathbb{K}_j$ with a statistical hypothesis H_j, by setting the credal set for the variables V to $\mathbb{K}_j$ for the valuation j of the hypothesis node H_j.

Output: a statistical model in which each hypothesis H_j is associated with a credal set over the variables V.

We can then use the common machinery to calculate the interval-valued probabilities that attach to the respective statistical hypotheses H_j.

We think that this can be useful to both statisticians and working scientists. It frequently happens that the bearing of some piece of evidence on the available hypotheses is somehow vague. In scientific investigation, data is often related to the hypothesis under scrutiny in a vague way. For example, we may have found an ancient coin on an archeological site and then wonder whether this finding is the result of a Roman settlement on that spot. We might think that it is more probable than not that such a coin is found if there has been a settlement. Yet it is not easy to attach a sharp probability value to that. Another case of vague evidential relations is presented by statistical hypotheses that allow for variation in the chances. For example, if a patient has a certain disease, a hypothesis may be that the probability of a positive test result may be between 60% and 80%. Typically, the hypothesis will be motivated by a number of studies, and the percentages within these studies may vary. In this way, interval-valued likelihoods present an interesting connection to evidential probability as well.

13.2.2 Logically Complex Statements with Statistical Hypotheses

The idea to include hypotheses in the credal network invites a further application that derives from the standard semantics.

Taken by themselves, second-order probabilities over statistical hypotheses allow one to infer various useful quantities, such as expectation values and predictions. By including the statistical hypotheses in the credal network, we can also infer probability assignments over logically complex propositions concerning statistical hypotheses, these inferred quantities, and the data. Importantly, it allows us to assess the logical combination of statistical hypotheses from analyses concerning overlapping sets of propositional variables.

As briefly discussed at the end of §6, we must be careful in interpreting the resulting probability assignments over logically complex propositions involving hypotheses, which are typically interval-valued. The interval-valued assignments to such propositions cannot be interpreted as credence intervals: they pertain to a single statistical hypothesis and not to a range of values of statistical parameters. The shape of the second-order probability, as introduced in the preceding section, tells us a lot more about the uncertainty over the probabilities assigned to variables, and such uncertainty over probabilities is simply not expressed in the interval probabilities of credal networks.

Statisticians may nevertheless make good use of credal networks that include statistical hypotheses. Often enough the proposition of interest in a scientific enquiry is logically complex, and very often the background knowledge contains logically complex propositions. With the above ideas in place, Schema (1.1) provides a systematic way of dealing with such pieces of background knowledge.

Chapter 14
Networks for Objective Bayesianism

According to the objective Bayesian semantics, $\varphi_1^{X_1}, \varphi_2^{X_2}, \ldots, \varphi_n^{X_n} \approx \psi^Y$ if and only if all the probability functions that satisfy the constraints imposed by the premisses and that are otherwise maximally equivocal, also satisfy the conclusion.

As outlined in §7, the left-hand side is interpreted as the set of constraints χ directly transferred from the agent's evidence $\mathcal{E}$. One then determines the set $\mathbb{P}_\chi^{\uplus}$ of probability functions that satisfy maximal consistent subsets of χ and one takes the convex closure $\left\langle \mathbb{P}_\chi^{\uplus} \right\rangle$ of this set to yield the set $\mathbb{E}$ of probability functions compatible with the agent's evidence $\mathcal{E}$. When premisses are regular $\mathbb{P}_\chi$ is already nonempty, closed and convex and $\mathbb{E} = \left\langle \mathbb{P}_\chi^{\uplus} \right\rangle = \mathbb{P}_\chi$. Hence under the assumption of regularity we can simply consider maximally equivocal probability functions from all those that satisfy the premisses.

14.1 Propositional Languages

Suppose further that $\varphi_1, \ldots, \varphi_n, \psi$ are sentences of a finite propositional language $\mathcal{L}$ with propositional variables $A_1, \ldots, A_n$. Then entropy is a measure of the degree to which a probability function equivocates and there is a unique entropy maximiser. In principle then we consider only a single model of the premisses, the probability function that has maximum entropy from all those that satisfy the premisses, and see whether that satisfies the conclusion. This model can be represented by a Bayesian network, and one can use this network to calculate the probability Y to attach to the conclusion sentence ψ. A Bayesian network that represents a probability function that represents degrees of belief that are deemed rational according to objective Bayesian epistemology is called an *objective Bayesian network* (Williamson, 2005b). In our scenario an objective Bayesian network can be constructed as follows.

Algorithm 12 Construction of an objective Bayesian network (i.e., a Bayesian network for the objective Bayesian semantics) on a propositional language.

R. Haenni et al., *Probabilistic Logics and Probabilistic Networks*, Synthese Library 350, DOI 10.1007/978-94-007-0008-6_14, © Springer Science+Business Media B.V. 2011

Input: $\varphi_1^{X_1}, \ldots, \varphi_n^{X_n}, \psi$ where $\varphi_1, \ldots, \varphi_n, \psi$ are sentences of a finite propositional language.

1. **Construction of the graph**:

   - Construct an undirected *constraint graph* $\mathcal{G}$ by taking the propositional variables occurring in $\varphi_1, \ldots, \varphi_n, \psi$ as nodes and connecting two variables with an edge if they occur in the same premise φ_i.
   - Transform $\mathcal{G}$ into a DAG $\mathcal{H}$:
     - Triangulate $\mathcal{G}$ to give $\mathcal{G}^T$.
     - Reorder the variables according to maximum cardinality search.
     - Let $D_1, \ldots, D_l$ be the cliques of $\mathcal{G}^T$, ordered according to highest labelled vertex.
     - Let $E_j = D_j \cap \left(\bigcup_{i=1}^{j-1} D_i \right)$ and $F_j = D_j \backslash E_j$, for $j = 1, \ldots, l$.
     - Construct a directed acyclic graph $\mathcal{H}$ by taking propositional variables as nodes, and
       - Add an arrow from each vertex in E_j to each vertex in F_j, for $j = 1, \ldots, l$.
       - Add further arrows, from lower numbered variables to higher numbered variables, to ensure that there is an arrow between each pair of vertices in $D_j, j = 1, \ldots, l,$.

2. **Derivation of the conditional probability functions**: Find the values of $P\left(a_i^e \mid par_i^e\right)$ that maximise the total entropy

$$H = -\sum_{i=1}^{n} \sum_{e \in \{0,1\}^{|Anc_i'|}} \left(\prod_{A_j \in Anc_i'} P\left(a_j^e \mid par_j^e\right) \right) \log P\left(a_i^e \mid par_i^e\right),$$

where Anc_i' is the set consisting of A_i and its ancestors in $\mathcal{H}$. (Here standard numerical techniques or Lagrange multiplier methods can be used.)

Output: $\mathcal{H}, \{P\left(a_i^e \mid par_i^e\right) : i = 1, \ldots, n, e \in \{0,1\}^n\}$

The output of this algorithm is provably a Bayesian network representation of the maximum entropy probability function P, from all those that satisfy the input (Williamson, 2005a, §5.7). This objective Bayesian network can be used to calculate $Y = P(\psi)$, as indicated in §8.

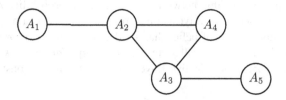

Fig. 14.1: Constraint graph.

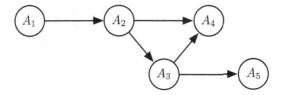

Fig. 14.2: Graph satisfying the Markov Condition.

Example 24. Suppose we have a question of the form:

$$a_1 \wedge \neg a_2^{[0.8,0.9]}, (\neg a_4 \vee a_3) \to a_2^{0.2}, a_5 \vee a_3^{[0.3,0.6]}, a_4^{0.7} \approx a_5 \to a_1?$$

This is short for the following question: given that $a_1 \wedge \neg a_2$ has probability between 0.8 and 0.9 inclusive, $(\neg a_4 \vee a_3) \to a_2$ has probability 0.2, $a_5 \vee a_3$ has probability in $[0.3, 0.6]$ and a_4 has probability 0.7, what probability should $a_5 \to a_1$ have? As explained in §7.3, this question can be given an objective Bayesian interpretation: supposing the agent's evidence imposes the constraints $P(a_1 \wedge \neg a_2) \in [0.8, 0.9], P((\neg a_4 \vee a_3) \to a_2) = 0.2, P(a_5 \vee a_3) \in [0.3, 0.6], P(a_4) = 0.7$, how strongly should she believe $a_5 \to a_1$? By means of Algorithm 12, an objective Bayesian network can be constructed to answer this question. First construct undirected constraint graph Figure 14.1 by linking variables that occur in the same constraint. Next, follow the algorithm to transform the undirected graph into a directed acyclic graph satisfying the Markov Condition, such as Figure 14.2. The third step is to maximise entropy to determine the probability distribution of each variable conditional on its parents in the directed graph. This yields the objective Bayesian network. Finally we use the network to calculate the probability of the conclusion

$$P(a_5 \to a_1) = P(\neg a_5 \wedge a_1) + P(a_5 \wedge a_1) + P(\neg a_5 \wedge \neg a_1)$$
$$= P(a_1) + P(\neg a_5|\neg a_1)(1 - P(a_1))$$

Thus we must calculate $P(a_1)$ and $P(\neg a_5|\neg a_1)$ from the network, which can be done using the common machinery of §8.2.

14.2 Predicate Languages

If $\varphi_1, \ldots, \varphi_n, \psi$ are sentences of a predicate language, a similar procedure can be applied.

The simplest case is that in which the constraints imposed by the left-hand side of the fundamental question, $\varphi_1^{X_1}, \varphi_2^{X_2}, \ldots, \varphi_n^{X_n}$, constrain finitely many atomic sentences (sentences of the form Ut). This occurs for example when the language is finite (i.e., there are finitely many constant, predicate, relation and function symbols), or when the sentences $\varphi_1, \ldots, \varphi_n, \psi$ are all quantifier-free. Let $a_1, a_2, \ldots, a_k$ be the atomic sentences constrained by $\varphi_1^{X_1}, \varphi_2^{X_2}, \ldots, \varphi_n^{X_n}$. Then we construct the objective Bayesian network exactly as in Algorithm 12, taking variables corresponding to these atomic sentences as nodes in the graph. (As pointed out in §7, in this finite case minimising distance to the equivocator is equivalent to maximising entropy.) The objective Bayesian network can then be used to calculate $P(\psi)$ as discussed in §8.

In the more general situation—with an infinite predicate language and arbitrary sentential constraints—steps need to be taken to ensure the probabilistic network remains finite (in order to use the network for computation). Moreover, $P \in \downarrow\mathbb{E}$ may not be a singleton; hence a credal network rather than a Bayesian network may be required to represent $\downarrow\mathbb{E}$. In this case we can use the following algorithm:[1]

Algorithm 13 Construction of an objective credal network on a predicate language.

Input: $\varphi_1^{X_1}, \ldots, \varphi_n^{X_n}, \psi$ where $\varphi_1, \ldots, \varphi_n, \psi$ are sentences of a predicate language and ψ is quantifier-free.

1. **Construction of the graph**:

   - Instantiate all quantified premise sentences. Let $t_1, \ldots, t_m$ be the constant symbols appearing in $\varphi_1, \ldots, \varphi_n, \psi$. For each occurrence of $\forall x \cdots (x) \cdots$ in the premisses substitute $\bigwedge_{i=1}^m \cdots t_i \cdots$. Similarly for each occurrence of $\exists x \cdots (x) \cdots$ in the premisses substitute $\bigvee_{i=1}^m \cdots t_i \cdots$.
   - Construct an undirected constraint graph $\mathcal{G}$:
     - Let $a_1, \ldots, a_k$ be the atomic sentences occurring in ψ and in the transformed premises. Construct binary variables $A_1, \ldots, A_k$ such that each A_i has a_i and $\neg a_i$ as possible assignments. Take these variables as the nodes of $\mathcal{G}$.
     - Connect two variables A_i and A_j with an edge if a_i and a_j occur in the same premise sentence φ_k.
   - Transform $\mathcal{G}$ into a DAG $\mathcal{H}$:
     - Triangulate $\mathcal{G}$ to give $\mathcal{G}^T$.
     - Reorder the variables according to maximum cardinality search.
     - Let $D_1, \ldots, D_l$ be the cliques of $\mathcal{G}^T$, ordered according to highest labelled vertex.
     - Let $E_j = D_j \cap (\bigcup_{i=1}^{j-1} D_i)$ and $F_j = D_j \backslash E_j$, for $j = 1, \ldots, l$.
     - Construct a directed acyclic graph $\mathcal{H}$ by taking variables $A_1, \ldots, A_K$ as nodes, and
       · Add an arrow from each vertex in E_j to each vertex in F_j, for $j = 1, \ldots, l$.
       · Add further arrows, from lower numbered variables to higher numbered variables, to ensure that there is an arrow between each pair of vertices in $D_j, j = 1, \ldots, l,$.

---

[1] See Williamson (2008a) for further discussion of this procedure.

2. **Derivation of the conditional probability functions**: Determine closed intervals for $P(a_i^e \mid par_i^e)$, where P ranges over those probability functions on the predicate language that satisfy the input constraints and are closest to the equivocator.

Output: a credal network on $\mathcal{H}$.

That the graph represents the independencies of any function P that satisfies the constraints and is closest to the equivocator follows for the same reasons as in the case of Algorithm 12 (Williamson, 2005a, §5.7). As in the case of Algorithm 12 we leave the methods for filling in the conditional probability intervals unspecified: what is important for the purposes of this book is the dimension reduction offered by representing P by a probabilistic network factorisation of P with respect to $\mathcal{H}$.

Fig. 14.3: A constraint graph and a resulting graph satisfying the Markov Condition.

Here is a simple example.

Example 25. Suppose we have a question of the form:

$$\forall x(Ux \to Vx)^{3/5}, \forall x(Vx \to Wx)^{3/4}, Ut^{[0.8,1]} \models Wt?$$

By means of Algorithm 13, an objective Bayesian network can be constructed to answer this question. There is only one constant symbol so instantiating the constraint sentences gives $Ut \to Vt, Vt \to Wt, Ut$. Let A_1 take assignments Ut (a_1) and $\neg Ut$ $(\bar{a}_1)$, A_2 take assignments Vt (a_2) and $\neg Vt$ $(\bar{a}_2)$ and A_3 take assignments Wt (a_3) and $\neg Wt$ $(\bar{a}_3)$. Then $\mathcal{G}$ is depicted on the left-hand side of Fig. 14.3 and $\mathcal{H}$ is depicted on the right-hand side of Fig. 14.3. It is not hard to see that $P(a_1) = 4/5, P(a_2|a_1) = 3/4, P(a_2|\neg a_1) = 1/2, P(a_3|a_2) = 5/6, P(a_3|\neg a_2) = 1/2$; together with $\mathcal{H}$, these probabilities yield a Bayesian network. The common machinery of §8.2 then gives us $P(a_3) = 11/15$ as an answer to our question.

Chapter 15
Conclusion

In this book we have argued in favour of two basic claims:

Part I: a unifying framework for probabilistic logic can be constructed around what we have called the *Fundamental Question of Probabilistic Logic*, or simply Schema (1.1);

Part II: probabilistic networks can provide a calculus for probabilistic logic—in particular they can be used to provide answers to the fundamental question.

These two claims constitute what we call the *progicnet* programme, and offer a means of unifying various approaches to combining probability and logic in a way that seems promising for practical applications.

Because of this twin focus and the programmatic nature of this book, we have neither been able to address all concerns about probabilistic logics nor address concerns about the various semantics discussed here. Many of these concerns (too many to list here!) are discussed in the supporting references. There are of course possible semantics for the fundamental question other than those considered here; we hope that this book will encourage research into how well these fit into the progicnet programme.

One locus for future research is the question of how answers to the fundamental question might influence decision making. Indeed the wider question of the relationship between probabilistic logic and decision theory has received scant attention in the literature. However this is clearly a crucial question both from a computational and from a philosophical point of view. If resources are bounded then certain queries in probabilistic logic will be most prudent as a basis for decision—but which? If these queries are to be used as a basis for decision then certain semantics may be more viable than others—but which?

R. Haenni et al., *Probabilistic Logics and Probabilistic Networks*, Synthese Library 350,   139
DOI 10.1007/978-94-007-0008-6_15, © Springer Science+Business Media B.V. 2011

References

Abraham, J. A. (1979). An improved algorithm for network reliability. *IEEE Transactions on Reliability*, 28:58–61.

Andréka, H., van Benthem, J., and Németi, I. (1998). Modal languages and bounded fragments of predicate logic. *Journal of Philosophical Logic*, 27:217–274.

Anrig, B. (2000). A generalization of the algorithm of Abraham. In Nikulin, M. and Limnios, N., editors, *MMR'2000: Second International Conference on Mathematical Methods in Reliability*, pages 95–98, Bordeaux, France.

Antonucci, A., Zaffalon, M., Ide, J. S., and Cozman, F. G. (2006). Binarization algorithms for approximate updating in credal nets. In Penserini, L., Peppas, P., and Perini, A., editors, *STAIRS'06, 3rd European Starting AI Researcher Symposium*, pages 120–131, Riva del Garda, Italy.

Arló-Costa, H. and Parikh, R. (2005). Conditional probability and defeasible inference. *Journal of Philosophical Logic*, 34:97–119.

Bacchus, F. A., Grove, D., Halpern, J., and Koller, D. (1993). Statistical foundations for default reasoning. In Bajcsv R., editor, *Proceedings of the International Joint Conference on Artificial Intelligence (IJCAI)*, pages 563–569. Chambéry, France.

Barnett, V. (1999). *Comparative Statistical Inference*. John Wiley, New York, NY.

Bernardo, J. M. and Smith, A. F. M. (2000). *Bayesian Theory*. Wiley, New York, NY.

Bernoulli, J. (1713). *Ars Conjectandi*. The Johns Hopkins University Press, Baltimore, 2006 edition. Trans. Edith Dudley Sylla.

Boutilier, C., Friedman, N., Goldszmidt, M., and Koller, D. (1996). Context-specific independence in Bayesian networks. In Horvitz, E. and Jensen, F., editors, *UAI'96, 12th Conference on Uncertainty in Artificial Inteligence*, pages 115–123, Portland, USA.

Burks, A. (1953). The presupposition theory of induction. *Philosophy of Science*, 20(3):177–197.

Cano, A., Fernández-Luna, J. M., and Moral, S. (2002). Computing probability intervals with simulated annealing and probability trees. *Journal of Applied Non-Classical Logics*, 12(2):151–171.

Cano, A. and Moral, S. (1996). A genetic algorithm to approximate convex sets of probabilities. In *IPMU'96, 6th international Conference on Information Process-*

ing and Management of Uncertainty in Knowledge-Based Systems, pages 859–864, Granada, Spain.

Carnap, R. (1950). *Logical Foundations of Probability*. University of Chicago Press, Chicago, IL.

Carnap, R. (1952). *The Continuum of Inductive Methods*. University of Chicago Press, Chicago, IL.

Carnap, R. (1962). *The Logical Foundations of Probability*. 2nd edition, University of Chicago Press, Chicago, IL.

Carnap, R. (1968). On rules of acceptance. In Lakatos, I., editor, *The Problem of Inductive Logic*, pages 146–150. North-Holland Co., Amsterdam, The Netherlands.

Carnap, R. (1980). A basic system of inductive logic, part ii. In Jeffrey, R., editor, *Studies in Inductive Logic and Probability*, volume 2, pages 7–150. University of California Press, Berkley, CA.

Chavira, M. and Darwiche, A. (2005). Compiling Bayesian networks with local structure. In *IJCAI'05, 19th International Joint Conference on Artificial Intelligence*, Edinburgh, U.K.

Chavira, M. and Darwiche, A. (2007). Compiling Bayesian networks using variable elimination. In *IJCAI'07, 20th International Joint Conference on Artificial Intelligence*, Hyderabad, India.

Cox, R. (1946). Probability, frequency and reasonable expectation. *American Journal of Physics*, 14(1):1–3.

Cozman, F. (forthcoming). Sets of probability distributions, independence, and convexity. *Synthese*.

Cozman, F., Haenni, R., Romeijn, J. W., Russo, F., Wheeler, G., and Williamson, J., editors (2008). *Combining Probability and Logic: Journal of Applied Logic*, 7(2).

Cozman, F. G. (2000). Credal networks. *Artificial Intelligence*, 120:199–233.

Cozman, F. G. and de Campos, C. P. (2004). Local computation in credal networks. In *ECAI'04, 16th European Conference on Artificial Intelligence, Workshop 22 on "Local Computation for Logics and Uncertainty"*, pages 5–11, Valencia, Spain.

da Rocha, J. C. F. and Cozman, F. G. (2003). Inference in credal networks with branch-and-bound algorithms. In Bernard, J. M., Seidenfeld, T., and Zaffalon, M., editors, *ISIPTA'03, 3rd International Symposium on Imprecise Probabilities and Their Applications*, pages 480–493, Lugano, Switzerland.

da Rocha, J. C. F., Cozman, F. G., and de Campos, C. P. (2003). Inference in polytrees with sets of probabilities. In Meek, C. and Kjærulff, U., editors, *UAI'03, 19th Conference on Uncertainty in Artificial Intelligence*, pages 217–224, Acapulco, Mexico.

Darwiche, A. (2002). A logical approach to factoring belief networks. In Fensel, D., Giunchiglia, F., McGuinness, D. L., and Williams, M. A., editors, *KR'02, 8th International Conference on Principles and Knowledge Representation and Reasoning*, pages 409–420, Toulouse, France.

Darwiche, A. and Marquis, P. (2002). A knowledge compilation map. *Journal of Artificial Intelligence Research*, 17:229–264.

Dawid, A. P. and Stone, M. (1982). The functional-model basis of fiducial inference (with discussion). *Annals of Statistics*, 10(4):1054–1074.

de Cooman, G. and Miranda, E. (2007). Symmetry of models versus models of symmetry. In Harper, W. L. and Wheeler, G. R., editors, *Probability and Inference: Essays in Honour of Henry E. Kyburg Jr.*, pages 67–149. College Publications, London.

de Finetti, B. (1937a). Foresight. its logical laws, its subjective sources. In Kyburg, H. E. and Smokler, H. E., editors, *Studies in Subjective Probability*, pages 53–118. Robert E. Krieger Publishing Company, Huntington, New York, NY second (1980) edition.

de Finetti, B. (1937b). La prévision: ses lois logiques, ses sources subjectives. *Annales de l'Institut Henri Poincaré*, 7(1):1–68.

de Finetti, B. (1974). *Theory of Probability: A Critical Introductory Treatment*. Wiley, New York, NY.

Dempster, A. (1963). On direct probabilities. *Journal of the Royal Statistical Society, Series B*, 35:100–110.

Dempster, A. P. (1968). A generalization of Bayesian inference. *Journal of the Royal Statistical Society*, 30:205–247.

Dias, P. M. C. and Shimony, A. (1981). A critique of Jaynes' maximum entropy principle. *Advances in Applied Mathematics*, 2(2):172–211.

Dubbins, L. E. (1975). Finitely additive conditional probability, conglomerability, and disintegrations. *Annals of Probability*, 3:89–99.

Dubois, D. and Prade, H. (1980). *Fuzzy Sets and Systems: Theory and Applications*. Kluwer, North Holland.

Earman, J. (1992). *Bayes or Bust?* MIT press, Cambridge, MA.

Fagin, R. and Halpern, J. Y. (1991). Uncertainty, belief, and probability. *Computational Intelligence*, 6:160–173.

Fagin, R. and Halpern, J. Y. (1994). Reasoning about knowledge and probability. *Journal of ACM*, 41(2):340–367.

Fagin, R., Halpern, J. Y., and Megiddo, N. (1990). A logic for reasoning about probabilities. *Information and Computation*, 87(1-2):78–128.

Fagin, R., Halpern, J. Y., Moses, Y., and Vardi, M. Y. (2003). *Reasoning About Knowledge*. MIT Press, Cambridge, MA.

Festa, R. (1993). *Optimum Inductive Methods*. Kluwer, Dordrecht.

Festa, R. (2006). Analogy and exchangeability in predictive inferences. *Erkenntnis*, 45:89–112.

Fisher, R. A. (1922). On the mathematical foundations of theoretical statistics. *Philosophical Transactions of the Royal Society of London*, 222:309–368.

Fisher, R. A. (1930). Inverse probability. *Proceedings of the Cambridge Philosophical Society*, 26:528–535.

Fisher, R. A. (1935). The fiducial argument in statistical inference. *Annals of Eugenics*, 6:317–324.

Fisher, R. A. (1936). Uncertain inference. *Proceedings of the American Academy of Arts and Sciences*, 71:245–258.

Fisher, R. A. (1956). *Statistical Methods and Scientific Inference*. Oliver and Boyd, Edinburgh.

Gaifman, H. and Snir, M. (1982). Probabilities over rich languages. *Journal of Symbolic Logic*, 47(3):495–548.

Galavotti, M. C. (2005). *A Philosophical Introduction to Probability*, volume 167 of *CSLI Lecture Notes*. Center for the Study of Language and Information.

Gelman, A., Carlin, J. B., Stern, H. S., and Rubin, D. B. (2003). *Bayesian Data Analysis*, 2nd edition, Chapman & Hall, Boca Raton, FL.

Good, I. J. (1965). *The Estimation of Probabilities*. MIT Press, Cambridge.

Hacking, I. (1965). *The Logic of Statistical Inference*. Cambridge University Press, Cambridge, MA.

Haenni, R. (2002). A query-driven anytime algorithm for argumentative and abductive reasoning. In Bustard, D., Liu, W., and Sterrit, R., editors, *Soft-Ware 2002, 1st International Conference on Computing in an Imperfect World*, LNCS 2311, pages 114–127. Springer, Belfast, UK.

Haenni, R. (2003). Anytime argumentative and abductive reasoning. *Soft Computing – A Fusion of Foundations, Methodologies and Applications*, 8(2):142–149.

Haenni, R. (2005a). Towards a unifying theory of logical and probabilistic reasoning. In Cozman, F. B., Nau, R., and Seidenfeld, T., editors, *ISIPTA'05, 4th International Symposium on Imprecise Probabilities and Their Applications*, pages 193–202, Pittsburgh, USA.

Haenni, R. (2005b). Using probabilistic argumentation for key validation in public-key cryptography. *International Journal of Approximate Reasoning*, 38(3):355–376.

Haenni, R. (2007). Climbing the hills of compiled credal networks. In de Cooman, G., Vejnarová, J., and Zaffalon, M., editors, *ISIPTA'07, 5th International Symposium on Imprecise Probabilities and Their Applications*, pages 213–222, Prague, Czech Republic.

Haenni, R. (2009). Probabilistic argumentation. *Journal of Applied Logic*, 7(2):155–176.

Haenni, R. and Hartmann, S. (2006). Modeling partially reliable information sources: a general approach based on Dempster-Shafer theory. *International Journal of Information Fusion*, 7(4):361–379.

Haenni, R., Kohlas, J., and Lehmann, N. (2000). Probabilistic argumentation systems. In Gabbay, D. M. and Smets, P., editors, *Handbook of Defeasible Reasoning and Uncertainty Management Systems*, volume 5: Algorithms for Uncertainty and Defeasible Reasoning, pages 221–288. Kluwer Academic Publishers, Dordrecht, Netherlands.

Haenni, R. and Lehmann, N. (2003). Probabilistic argumentation systems: a new perspective on Dempster-Shafer theory. *International Journal of Intelligent Systems, Special Issue on the Dempster-Shafer Theory of Evidence*, 18(1):93–106.

Haenni, R., Romeijn, J., Wheeler, G., and Williamson, J. (2008). Possible semantics for a common framework of probabilistic logics. In Huynh, V. N., editor, *UncLog'08, International Workshop on Interval/Probabilistic Uncertainty and Non-Classical Logics*, Advances in Soft Computing, Ishikawa, Japan.

Hailperin, T. (1996). *Sentential Probability Logic*. Lehigh University Press, Bethlehem, PA.

Halmos, P. R. (1950). *Measure Theory*. Van Nostrand Reinhold Company, New York, NY.

Halpern, J. and Pucella, R. (2005). Evidence with uncertain likelihoods. *UAI '05, Proceedings of 21st Conference on Uncertainty in Artificial Intelligence*, pages 243–250. Edinburgh, Scotland.

Halpern, J. Y. (1990). An analysis of first-order logics of probability. *Artificial Intelligence*, 46:311–350.

Halpern, J. Y. (2003). *Reasoning about Uncertainty*. MIT Press, Cambridge, MA.

Halpern, J. Y. and Fagin, R. (1992). Two views of belief: belief as generalized probability and belief as evidence. *Artificial Intelligence*, 54(3):275–317.

Harper, W. and Wheeler, G., editors (2007). *Probability and Inference: Essays In Honor of Henry E. Kyburg, Jr.* College Publications, London.

Harper, W. L. (1981). Kyburg on direct inference. In Bogdan, R., editor, *Profile of Kyburg and Levi*, pages 97–127. Dordrecht, Reidel.

Hawthorne, J. and Makinson, D. C. (2007). The quantitative/qualitative watershed for rules of uncertain inference. *Studia Logica*, 86:247–297.

Horn, A. and Tarski, A. (1948). Measures in Boolean Algebras. *Transactions of the AMS*, 64(1): 467–497.

Howson, C. (2001). The logic of Bayesian probability. In Corfield, D. and Williamson, J., editors, *Foundations of Bayesianism*, Applied Logic Series, pages 137–159. Kluwer Academic Publishers, Dordrecht, Netherlands.

Howson, C. (2003). Probability and logic. *Journal of Applied Logic*, 1(3-4):151–165.

Howson, C. and Urbach, P. (1993). *Scientific Reasoning: the Bayesian aproach*. Open Court Publishing Company, Chicago, IL.

Ide, J. S. and Cozman, F. G. (2004). IPE and L2U: Approximate algorithms for credal networks. In *STAIRS'04, 2nd European Starting AI Researcher Symposium*, pages 118–127, Valencia, Spain.

Ide, J. S. and Cozman, F. G. (2005). Approximate inference in credal networks by variational mean field methods. In Cozman, F. G., Nau, R., and Seidenfeld, T., editors, *ISIPTA'05, 4th International Symposium on Imprecise Probabilities and Their Applications*, pages 203–212, Pittsburgh, USA.

Jaeger, M. (2002). Relational bayesian networks: a survey. *Electronic Transactions in Artificial Intelligence*, 6.

Jaynes, E. T. (1957). Information theory and statistical mechanics. *The Physical Review*, 106(4):620–630.

Jaynes, E. T. (2003). *Probability Theory: The Logic of Science*. Cambridge University Press, Cambridge, MA.

Jeffrey, R. (1956). Valuation and acceptance of scientific hypotheses. *Philosophy of Science*, 23(3):237–246.

Jeffrey, R. (1965). *Logic of Decision*. McGraw-Hill, New York, NY.

Jeffrey, R. (1992). *Probability and the Art of Judgment*. Cambridge University Press, Cambridge, MA.

Jøsang, A. (1997). Artificial reasoning with subjective logic. In Nayak, A. C. and Pagnucco, M., editors, *2nd Australian Workshop on Commonsense Reasoning*, Perth, Australia.

Kersting, K. and Raedt, L. D. (2007). Bayesian logic programming: Theory and tool. In Getoor, L. and Taskar, B., editors, *Introduction to Statistical Relational Learning*. MIT Press, Cambridge, MA.

Kersting, K., Raedt, L. D., and Raiko, T. (2006). Logical hidden markov models. *Journal of Artificial Intelligence Research*, 25:425–456.

Kohlas, J. (2003). Probabilistic argumentation systems: A new way to combine logic with probability. *Journal of Applied Logic*, 1(3–4):225–253.

Kohlas, J. and Monney, P. A. (2008). *Statistical Information: Assumption-Based Statistical Inference*, volume 3 of *Sigma Series in Stochastics*. Heldermann Verlag, Lemgo, Germany.

Kohlas, J. and Shenoy, P. P. (2000). Computation in valuation algebras. In Gabbay, D. M. and Smets, P., editors, *Handbook of Defeasible Reasoning and Uncertainty Management Systems*, volume 5: Algorithms for Uncertainty and Defeasible Reasoning, pages 5–39. Kluwer Academic Publishers, Dordrecht, Netherlands.

Kolmogorov, A. N. (1950). *Foundations of the Theory of Probability*. Chelsea Publishing Company, New York, USA.

Kraus, S., Lehman, D., and Magidor, M. (1990a). Nonmonotonic reasoning, preferential models and cumulative logics. *Artificial Intelligence*, 44:167–207.

Kraus, S., Lehmann, D., and Magidor, M. (1990b). Nonmonotonic reasoning, preferential models and cumulative logics. *Artificial Intelligence*, 44:167–207.

Kuipers, T. (1978). *Studies in Inductive Probability and Rational Expectation*. Reidel, Dordrecht.

Kyburg, Jr., H. E. (1961). *Probability and the Logic of Rational Belief*. Wesleyan University Press, Middletown, CT.

Kyburg, Jr., H. E. (1974). *The Logical Foundations of Statistical Inference*. D. Reidel, Dordrecht.

Kyburg, Jr., H. E. (1987). Bayesian and non-Bayesian evidential updating. *Artificial Intelligence*, 31:271–294.

Kyburg, Jr., H. E. (2003). Are there degrees of belief? *Journal of Applied Logic*, 1:139–149.

Kyburg, Jr., H. E. (2007). Bayesian inference with evidential probability. In Harper, W. and Wheeler, G., editors, *Probability and Inference: Essays in Honor of Henry E. Kyburg, Jr.*, pages 281–296. College Publications, London.

Kyburg, Jr., H. E. and Pittarelli, M. (1996). Set-based Bayesianism. *IEEE Transactions on Systems, Man and Cybernetics*, 26(3):324–339.

Kyburg, Jr., H. E. and Teng, C. M. (1999). Statistical inference as default logic. *International Journal of Pattern Recognition and Artificial Intelligence*, 13(2):267–283.

Kyburg, Jr., H. E. and Teng, C. M. (2001). *Uncertain Inference*. Cambridge University Press, Cambridge, MA.

Kyburg, Jr., H. E. and Teng, C. M. (2002). The logic of risky knowledge. In *Electronic Notes in Theoretical Computer Science*, volume 67. Elsevier Science, Amsterdam, The Netherlands.

Kyburg, Jr., H. E., Teng, C. M., and Wheeler, G. (2007). Conditionals and consequences. *Journal of Applied Logic*, 5(4):638–650.

Landwehr, N., Kersting, K., and Raedt, L. D. (2007). Integrating naïve Bayes and FOIL. *Journal of Machine Learning Research*, 8:481–507.

Laskey, K. B. (2008). MEBN: A language for first-order Bayesian knowledge bases. *Artificial Intelligence*, 172:140–178.

Lauritzen, S. and Spiegelhalter, D. J. (1988). Local computations with probabilities on graphical structures and their application to expert systems. *Journal of the Royal Statistical Society*, 50(2):157–224.

Lehman, D. and Magidor, M. (1990). What does a conditional knowledge base entail? *Artificial Intelligence*, 55:1–60.

Leuridan, B. (2008). *Laws of Nature and Causality in the Special Sciences: A Philosophical and Formal Analysis*. PhD thesis, University of Ghent, Belgium.

Levi, I. (1977). Direct inference. *Journal of Philosophy*, 74:5–29.

Levi, I. (1980). *The Enterprise of Knowledge: An Essay on Knowledge, Credal Probability, and Chance*. MIT Press, Cambridge, MA.

Levi, I. (2007). Probability logic and logical probability. In Harper, W. L. and Wheeler, G., editors, *Probability and Inference: Essays In Honor of Henry E. Kyburg, Jr.*, pages 255–266. College Publications, London.

Lindley, D. (1958). Fiducial distributions and Bayes' theorem. *Journal of the Royal Statistical Society, Series B*, 20:102–107.

Lindley, D. (1963). Inferences from multinomial data: Learning about a bag of marbles. *Journal of the Royal Statistical Society, Series B*, 58:3–57.

Makinson, D. (2005). *Bridges from Classical to Nonmonotonic Logic*. King's College Publications, London.

Marklund, S., Lijas, J., Rodriguez-Martinez, H., Rönnstrand, L., Funa, K., Moller, M., Lange, D., Edfors-Lilja, I., and Andersson, L. (1998). Molecular basis for the dominant white phenotype in the domestic pig. *Genome Research*, 8:826–833.

Mayo, D. (1996). *Error and the Growth of Experimental Knowledge*. University of Chicago Press, Chicago, IL.

Mood, A., Graybill, F., and Boes, D. (1974). *Introduction to the Theory of Statistics* 3rd edition. McGraw-Hill, New York, NY.

Morgan, C. G. (2000). The nature of nonmonotonic reasoning. *Minds and Machines*, 10:321–360.

Neapolitan, R. E. (1990). *Probabilistic Reasoning in Expert Systems*. Wiley, New York, NY.

Neapolitan, R. E. (2003). *Learning Bayesian Networks*. Prentice Hall, Upper Saddle River.

Neyman, J. (1957). Inductive behavior as a concept of philosophy of science. *Review of the International Statistical Institute*, 25:5–22.

Neyman, J. and Pearson, E. (1967). *Joint Statistical Papers*. University of California Press, Berkeley, CA.

Ngo, L. and Haddawy, P. (1995). Probabilistic logic programming and Bayesian networks. In *ACSC '95: Proceedings of the 1995 Asian Computing Science Conference on Algorithms, Concurrency and Knowledge*, pages 286–300, Springer London, UK.

Nilsson, N. J. (1986). Probabilistic logic. *Artificial Intelligence*, 28:71–87.

Nix, C. and Paris, J. (2007). A note on binary inductive logic. *Journal of Philosophical Logic*, 36(6):735–771.

Paris, J. and Simmonds, R. (2009). O is not enough. *The Review of Symbolic Logic*, 2:298–309.

Paris, J. B. (1994). *The Uncertain Reasoner's Companion*. Cambridge University Press, Cambridge, MA.

Pawlak, Z. (1991). *Rough Sets: Theoretical Aspects of Reasoning about Data*. Kluwer, Dordrecht.

Pearl, J. (1988). *Probabilistic Reasoning in Intelligent Systems*. Morgan Kaufmann, San Francisco, CA.

Pearl, J. (1990a). Reasoning with belief functions: An analysis of compatibility. *International Journal of Approximate Reasoning*, 4(5–6):363–389.

Pearl, J. (1990b). System Z: A natural ordering of defaults with tractable applications to default reasoning. In *Theoretical Aspects of Reasoning about Knowedge*, pages 121–135. Morgan Kaufmann Publishers Inc., San Francisco, CA, USA.

Pearl, J. (2000). *Causality: Models, Reasoning and Inference*. Cambridge University Press, New York, NY.

Pollock, J. (1993). Justification and defeat. *Artificial Intelligence*, 67:377–407.

Poole, D. (1993). Probabilistic Horn abduction and Bayesian networks. *Artificial Intelligence*, 64:81–129.

Press, J. (2003). *Subjective and Objective Bayesian Statistics: Principles, Models, and Applications*. John Wiley, New York, NY.

Ramsey, F. P. (1926). Truth and probability. In Kyburg, H. E. and Smokler, H. E., editors, *Studies in Subjective Probability*, pages 23–52. Robert E. Krieger Publishing Company, Huntington, New York, NY second (1980) edition.

Reichenbach, H. (1949). *The Theory of Probability*. University of California Press, Berkeley, CA.

Richardson, M. and Domingos, P. (2006). Markov logic networks. *Machine Learning*, 62:107–136.

Romeijn, J. (2005). *Bayesian Inductive Logic*. PhD dissertation, University of Groningen, The Netherland.

Romeijn, J. (2006). Analogical predictions for explicit similarity. *Erkenntnis*, 64:253–280.

Romeijn, J.-W., Haenni, R., Wheeler, G., and Williamson, J. (2009). Logical relations in a statistical problem, In Löwe, B.; Pacuit, E., and Romeijn, J.-W., editors, *Foundations of the Formal Sciences VI, Reasoning about Probabilities and Probabilistic Reasoning*, pages 49–79. College Publications, London.

Rosenkrantz, R. D. (1977). *Inference, Method and Decision: Towards a Bayesian Philosophy of Science*. Reidel, Dordrecht.

Ruspini, E. H. (1986). The logical foundations of evidential reasoning. Technical Report 408, SRI International, AI Center, Menlo Park, USA.

Ruspini, E. H., Lowrance, J., and Strat, T. (1992). Understanding evidential reasoning. *International Journal of Approximate Reasoning*, 6(3):401–424.

Russell, S. and Norvig, P. (2003). *Artificial Intelligence: A Modern Approach*. 2nd edition. Prentice Hall, Indiana, USA.

Schervish, M. J., Seidenfeld, T., Kadane, J. B., and Levi, I. (2003). Extensions of expected utility theory and some limitations of pairwise comparisons. In Bernard, J.-M., Seidenfeld, T., and Zaffalon, M., editors, *ISIPTA '03, Proceedings of the 3rd Intenational Symposium on Imprecise Probabilities and Their Applications*, pages 496–510. Lugano, Switzerland. Proceedings in informatics 18 Carleton Scientific 2003.

Seidenfeld, T. (1979). *Philosophical Problems of Statistical Inference: Learning from R.A. Fisher*. Reidel, Dordrecht.

Seidenfeld, T. (1992). R. A. Fisher's fiducial argument and Bayes' theorem. *Statistical Science*, 7:358–368.

Seidenfeld, T. (2007). Forbidden fruit: When Epistemic Probability may not take a bite of the Bayesian apple. In Harper, W. and Wheeler, G., editors, *Probability and Inference: Essays in Honor of Henry E. Kyburg, Jr.*, pages 267–279. College Publications, London.

Seidenfeld, T., Schervish, M. J., and Kadane, J. B. (2007). Coherent choice functions under uncertainty. In *Proceedings of the Fifth International Symposium on Imprecise Probability: Theories and Applications (ISIPTA)*, Prague, Czech Republic.

Seidenfeld, T. and Wasserman, L. (1993). Dilation for sets of probabilities. *The Annals of Statistics*, 21:1139–154.

Shafer, G. (1976). *A Mathematical Theory of Evidence*. Princeton University Press, Princeton, NJ.

Shenoy, P. P. and Shafer, G. (1988). Axioms for probability and belief-function propagation. In Shachter, R. D., Levitt, T. S., Lemmer, J. F., and Kanal, L. N., editors, *UAI'88, 4th Conference on Uncertainty in Artificial Intelligence*, pages 169–198, Minneapolis, USA.

Skyrms, B. (1993). Analogy by similarity in hyper-Carnapian inductive logic. In Earman, J., Janis, A. I., Massey, G., and Rescher, N., editors, *Philosophical Problems of the Internal and External Worlds*, pages 273–282. University of Pittsburgh Press, Pittsburgh, CA.

Smets, P. and Kennes, R. (1994). The transferable belief model. *Artificial Intelligence*, 66:191–234.

Spirtes, P., Glymour, C., and Scheines, R. (1993). *Causation, Prediction, and Search*. Springer, New York, NY.

Swift, T. and Wheeler, G. Extending description logics to support statistical information. *International Journal of Approximate Reasoning* (To appear).

Teng, C. M. (2007). Conflict and consistency. In Harper, W. L. and Wheeler, G., editors, *Probability and Inference: Essays in Honor of Henry E. Kyburg, Jr.*, pages 53–66. College Publications, London.

Thrun, S., Burgard, W., and Fox, D. (2005). *Probabilistic Robotics*. MIT Press, Cambridge, MA.

Wachter, M. and Haenni, R. (2006a). Logical compilation of Bayesian networks. Technical Report iam-06-006, University of Bern, Switzerland.

Wachter, M. and Haenni, R. (2006b). Propositional DAGs: a new graph-based language for representing Boolean functions. In Doherty, P., Mylopoulos, J., and Welty, C., editors, *KR'06, 10th International Conference on Principles of Knowledge Representation and Reasoning*, pages 277–285, U.K. AAAI Press, Lake District.

Wachter, M. and Haenni, R. (2007). Logical compilation of Bayesian networks with discrete variables. In Mellouli, K., editor, *ECSQARU'07, 9th European Conference on Symbolic and Quantitative Approaches to Reasoning under Uncertainty*, LNAI 4724, pages 536–547, Hammamet, Tunisia.

Wachter, M., Haenni, R., and Pouly, M. (2007). Optimizing inference in Bayesian networks and semiring valuation algebras. In Gelbukh, A. and Kuri Morales, A. F., editors, *MICAI'07: 6th Mexican International Conference on Artificial Intelligence*, LNAI 4827, pages 236–247, Aguascalientes, Mexico.

Walley, P. (1991). *Statistical Reasoning with Imprecise Probabilities*. Chapman and Hall, London.

Wheeler, G. (2004). A resource bounded default logic. In Delgrande, J. and Schaub, T., editors, *NMR 2004*, pages 416–422, Whistler, Canada.

Wheeler, G. (2006). Rational acceptance and conjunctive/disjunctive absorption. *Journal of Logic, Language and Information*, 15(1-2):49–63.

Wheeler, G. (2007). Two puzzles concerning measures of uncertainty and the positive Boolean connectives. In Neves, J., Santos, M., and Machado, J., editors, *The Proceedings of the 13th Portuguese Conference on Artificial Intelligence (EPIA 2007)*, LNAI, pages 170–180, Springer Berlin.

Wheeler, G. and Damásio, C. (2004). An implementation of statistical default logic. In Alferes, J. and Leite, J., editors, *Logics in Artificial Intelligence*, Lecture Notes in Artificial Intelligence, pages 121–133, Springer Berlin.

Wheeler, G. and Pereira, L. M. (2004). Epistemology and artificial intelligence. *Journal of Applied Logic*, 2(4):469–493.

Wheeler, G. and Williamson, J. (2010). Evidential probability and objective Bayesian epistemology. In Bandyopadhyay, P.S. and Forster, M., editors, *Hand-Book of the Philosophy of Statistics*. Elsevier, North Holland.

Williams, R. (2002). Algorithms for quantified Boolean formulas. In *SODA'02, 13th Annual ACM-SIAM Symposium on Discrete Algorithms*, pages 299–307, San Francisco, USA.

Williamson, J. (2002). Probability logic. In Gabbay, D., Johnson, R., Ohlbach, H. J., and Woods, J., editors, *Handbook of the Logic of Argument and Inference: The Turn Toward the Practical*, pages 397–424. Elsevier, Amsterdam, The Netherland.

Williamson, J. (2005a). *Bayesian Nets and Causality: Philosophical and Computational Foundations*. Oxford University Press, Oxford.

Williamson, J. (2005b). Objective Bayesian nets. In Artemov, S., Barringer, H., d'Avila Garcez, A. S., Lamb, L. C., and Woods, J., editors, *We Will Show Them! Essays in Honour of Dov Gabbay*, volume 2, pages 713–730. College Publications, London.

Williamson, J. (2007a). Inductive influence. *British Journal for the Philosophy of Science*, 58(4):689–708.

Williamson, J. (2007b). Motivating objective Bayesianism: from empirical constraints to objective probabilities. In Harper, W. L. and Wheeler, G. R., editors, *Probability and Inference: Essays in Honour of Henry E. Kyburg Jr.*, pages 151–179. College Publications, London.

Williamson, J. (2008a). Objective Bayesian probabilistic logic. *Journal of Algorithms in Cognition, Informatics and Logic*, 63:167–183.

Williamson, J. (2008b). Objective Bayesianism with predicate languages. *Synthese*, 163(3):341–356.

Williamson, J. (2009a). Objective Bayesianism, Bayesian conditionalisation and voluntarism. *Synthese*, doi 10.1007/s11229-009-9515-y.

Williamson, J. (2009b). Philosophies of probability. In Irvine, A., editor, *Handbook of the Philosophy of Mathematics*, pages 493–533. North-Holland, Amsterdam. Handbook of the Philosophy of Science volume 4.

Williamson, J. (2010). *In Defence of Objective Bayesianism*. Oxford University Press, Oxford.

Wiseman, J. (1986). *A History of the British Pig*. Ebenezer Baylis & Sons, Worcester.

Zadeh, L. A. (1979). On the validity of Dempster's rule of combination of evidence. Technical Report 79/24, University of California, Berkely, CA.

Index